# OXFORD STATISTICAL SCIENCE SERIES

SERIES EDITORS

J. B. COPAS   A. P. DAWID

G.K.EAGLESON   D.A.PIERCE   B.W.SILVERMAN

# OXFORD STATISTICAL SCIENCE SERIES

# Coordinate-Free Multivariable Statistics

## An Illustrated Geometric Progression from Halmos to Gauss and Bayes

MERVYN STONE

*University College London*

CLARENDON PRESS · OXFORD
1987

Oxford University Press, Walton Street, Oxford OX2 6DP

Oxford New York Toronto
Delhi Bombay Calcutta Madras Karachi
Petaling Jaya Singapore Hong Kong Tokyo
Nairobi Dar es Salaam Cape Town
Melbourne Auckland

and associated companies in
Beirut Berlin Ibadan Nicosia

Oxford is a trade mark of Oxford University Press

Published in the United States
by Oxford University Press, New York

British Library Cataloguing in Publication Data
Stone, Mervyn
Coordinate-free multivariable statistics:
an illustrated geometric progression from
Halmos to Gauss and Bayes.—(Oxford
statistical science series; 2).
1. Multivariate analysis
I. Title
519.5        QA278
ISBN 0-19-852210-X

Library of Congress Cataloguing in Publication Data
Stone, Mervyn.
Coordinate-free multivariable statistics.
(Oxford statistical science series)
Bibliography: p.
Includes index.
1. Multivariate analysis.   2. Vector spaces.
3. Transformations (Mathematics) I. Title. II. Series.
QA278.S76 1987     519.5'35        86-33205
ISBN 0-19-852210-X

Typeset and printed in Northern Ireland by
The Universities Press (Belfast) Ltd

# Preface

This book has only one objective, and that is to expose the simplicity and power of the coordinate-free, geometric approach as a theoretical framework for linear and affine multivariate methods in statistics. The few tools that we will need for this are all standard items in undergraduate-level linear algebra:

*real, finite-dimensional vector spaces and their duals*
*linear transformations and their duals*
*inner products and their duals*
*orthogonal projections.*

With these ingredients, a few derived notions, and some specially formulated theorems, there is developed a fairly conventional sequence of statistical problems and their optimal solution.

Any claims to novelty in the treatment rely on the combination of three features:

(i)   the extensive use of dual vector spaces
(ii)  the exploitation of statistically felicitous notation for certain linear and bilinear operators
(iii) the liberal admixture of revealing pictures.

It is, of course, the reader's privilege to assess the strength of these claims, and to judge whether valuable insights are indeed generated by the approach that gradually emerges. That opinions will undoubtedly differ is evidenced by a distinguished statistician's review (*Math. Rev.* 81d: 62064) of one of the papers in our list of references. The reviewer comments that the author 'has confused the issues by introducing his own notation and trying to interpret well-known and easily established results in linear estimation through complicated geometric figures'.

The dependence of this book on Halmos's classic work *Finite-dimensional vector spaces* should be apparent, at least in the extensive borrowings from it in matters of terminology and layout. For statistical lineage, I am happy to acknowledge the influence of the work of A. P. Dempster, M. L. Eaton, and W. H. Kruskal. Less remotely, my warmest thanks must go to A. P. Dawid for his generous encouragement and commentary, extending over a decade and a half, to Nadine Schuster for her efficient and accommodating word-processing, and to Wang Jinglong for his critical and constructive reading of the text and checking of exercises.

The crucially important pictures have emerged as a hybrid of line structures gracefully handcrafted by Wang Jinglong and lettering from OUP's elegant stable. This lettering was used merely to ensure an unambiguous line of communication with the text, and should not deter readers from the recommended do-it-yourself activity.

The exercises are designed to play a role as valuable as that of the illustrations. The reader who attempts more than a handful may, I hope, become addicted to the coordinate-free approach, and discover that any initial difficulties are more a matter of novelty than substance. However, it is necessarily the case that those already acquainted with the primitive concepts of the general linear model, least-squares estimation and prediction, and multivariate analysis will find the going easier. Mathematicians unfamiliar with the underlying statistical ideas could usefully, and perhaps amusingly, exchange familiarities with statisticians ignorant of the relevant linear algebra.

Finally, that this book is unlikely to be free of error is not just a reflection of its statistical status: I would welcome readers' corrections and suggestions for improvement.

*June 1986*                                                        M. S.
*London*

# Contents

# Preamble

If challenged, most teachers of mathematics would probably admit to having experienced, at one time or another, a singular embarrassment at the chalkface. In an attempt to portray some 3-dimensional object on the surface at hand, the teacher boldly draws a line that he sees as going, unambiguously, into the wall—only later to find that many of the students have, with equal justification, seen it doing something quite different. Again, in some ambitious representation of $p$ dimensions, a line and a flat that should never meet will insist on doing so, with adverse consequences for the teacher's reputation, if not for the advancement of mathematical science.

In the light of such difficulties, it is not at all surprising that authors of linear algebra textbooks and the like have, with few exceptions, eschewed pictorial representations of anything but the trivial. This rejection is all the more striking, since the textual treatments are often essentially geometrical and even go so far as to make use of 'word-pictures'. Halmos (1958, p. 6), commenting on vector spaces over the complex field, regrets that their

one great disadvantage is the difficulty of drawing pictures

while, elsewhere (p. 73), he advises his readers how to draw a

pair of axes (linear manifolds) in the plane (their direct sum). To make the picture look general enough, do not draw perpendicular axes!.

Kruskal (1975) also writes frankly about the problem of drawing pictures, and his remarks are worth quoting at length:

Another way of gaining a sense of concreteness is to draw balloon-shaped objects denoting the various linear manifolds we shall be considering . . . and to show on the resulting diagram arrows for [linear transformations and their inverses]. I understand that such aids to the mind are at best activities of child-like naiveté, akin to rhythmic toe tapping by a string quartet player, and with analogous dangers of misinterpretation and social contempt. Nonetheless, used quietly and with care not to substitute a highly special case for a proof, diagrams like the one suggested can be most useful; it would be disingenuous to remain silent about their utility.

Neither of these authors uses pictures in the locations cited. Although

Drygas (1970) does use pictures on the approaches to the core of his work, even he appears to exclude them when the proving gets tough.

One argument that has been adduced for this reticence in artistic display is that whatever value pictorial representation may have derives mainly from the individual act of encoding that it involves. A picture is necessarily built, component by component, in step with the accretion of understanding, and its final form will reflect the personal whims and needs of its architect. The mere inspection of the work of another mathematical artist, however well-executed it may by, cannot reveal the necessary dynamics and distinctions, and is no substitute for the do-it-yourself approach.

I trust there is enough falsity in this argument to justify the inclusion of more than 50 pictures in the chapters which follow. In any case, the reader always has the option of covering up any or all of these, and substituting his or her own. Part of the problem of drawing figures that guide and inform others is the absence of any widely-agreed conventions for the portrayal of the necessary components. It is the aim of the following notes to try to fill that gap, at least for present purposes. These notes should eliminate the grosser forms of confusion already alluded to. I hope they will also make the figures as 'readable', it not more so, than the algebraic proofs that in many cases they effectively replace.

○ Starting with one of Kruskal's 'balloon-shaped objects', Fig. 1 represents the vector space whose name, 'n' for short, is inserted *in script capitals* on its perimeter, which may be considered to be at infinity.

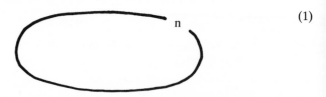

(1)

○ At the next level of complexity, n1 in Fig. 2 would be a vector space, and n2 would be an affine manifold (*flat*) of lower dimension than n1, either a subspace of n1 or an affine offset of a subspace of n1.

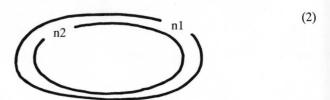

(2)

○ Points drawn within such perimeters usually represent single vectors belonging to the set defined by the perimeter. Thus, in Fig. 3, n2 must be a vector space, since it contains the origin (zero vector) 0. Singleton vectors will be given names *in lower case Roman*.

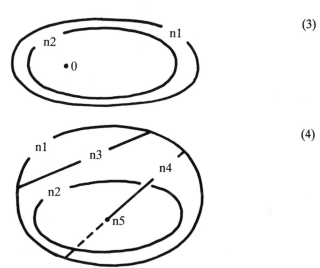

(3)

(4)

○ When there is no need to display any structure within a particular flat, it is prudent, in terms of expenditure of dimensions, to represent it as a straight line (which can be thought of as a squeezed enclosure). Thus, in Fig. 4, n3 is a flat that does not intersect the flat n2, whereas n4 is a flat whose intersection with n2 is n5. Drawn as a point, n5 would usually be a single vector but, if the representation were crowded, the point might be obliged to represent a flat. (The letter-type of the name would distinguish these cases.) Usually, a non-singleton intersection of two flats would be portrayed as in Fig. 5. The *broken* lines in (4) and (5) are inserted merely to enhance readability by an induced sense of perspective.

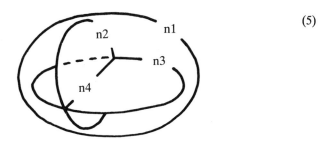

(5)

○ Already, an ambiguity needs to be resolved in connection with Fig. 4. Our convention is that, since no intersections are indicated, n3 intersects neither n2 nor n4. However, it is not clear whether or not n3 is contained in the span, ⟨n2, n4⟩, of n2 and n4, and, if it is not, whether or not it has any intersection with ⟨n2, n4⟩. (If n2 and n4 are vector spaces then ⟨n2, n4⟩ = n2 + n4.) The three possibilities are represented in Figs 6, 7, and 8, respectively. In (7), n6 (=n3 ∩ ⟨n2, n4⟩) may be more than a singleton, while the *dotted* line serves to remind us that n3 lives, apart from its point(s) n6, in higher dimensions than the representational 3 dimensions that have been occupied by n2 and n4. It is clear that, with pictures such as (7), we are close to the limits of feasible representation. It is fortunate that no higher level of complexity is needed to deal with the theory to be illustrated.

(6)

(7)

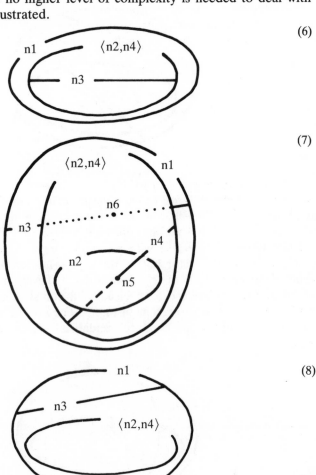

(8)

(9)

○ Sometimes, a vector needs to be shown in its free form, i.e. not tied down to the origin 0. Our convention is the standard one shown in Fig. 9: the vector n4 (lower case Roman) is the vector n3 minus the vector n2. Fig. 9 may also be used to illustrate *linear transformation,* in which case n4 would be the name, written in *capital* Roman, of a linear transformation taking n2 into n3. Such transformations may be shown acting between flats, as, for example, in Fig. 10, in which n1 is the domain, and n3 the range, of the transformation n4.

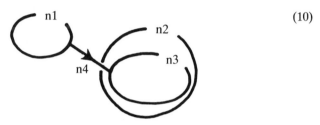

(10)

○ Our two remaining conventions concern the important concepts of *parallelism* and *perpendicularity* (orthogonality with respect to some inner product). The parallelism convention is in the spirit of Halmos's advice about perpendicularity, already cited: if two flats (for example n2 and n3 in Fig. 11) are drawn to *look* parallel then they *are* parallel. For orthogonality, we call in aid the symbol □, which may be drawn obliquely. This may represent either an ordinary right-angle between two vectors (e.g. n2, n3 in Fig. 12) or, in the role of freely rotating buttress, the orthogonality of a vector to the whole of a flat (e.g. n4 and n5).

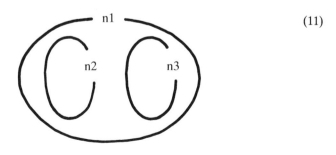

(11)

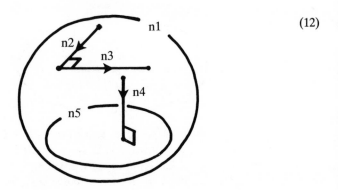

(12)

The name, or other distinguishing mark, of the inner product may be placed inside the □ symbol, when distinctions are called for.

# 1
# Spaces

This chapter introduces the basic ingredients of our coordinate-free approach to linear multivariate analysis, and outlines the kind of statistical problem that the fuller development in later chapters will be required to handle.

## §1. Statistical data

It is frequently the case that statistical data take the form of the observed, real *values* of a number of *variables* measured on each of a set of units. 'Units' might be individual human beings in one study, trials of a biochemical experiment in another, or just pieces of rock in a third. 'Variables' are designators such as 'weight in lbs', 'initial velocity of reaction in mols per second', or 'percentage of rock-type A'. 'Values' are the specific numerical records or measurements, corresponding to the conjunction of units and their designated variables. Thus, in the first study, the unit 'Tom' may be found to have the value 129 for the variable 'weight in lbs'.

In the subcategory of cases where linear methods are deemed appropriate, such methods may take a variety of forms.

(i)  When interest is focused on just one of the variables, the observed values for this variable may be averaged over units, perhaps after linear adjustment involving the observed values of the other variables.

(ii)  A different sort of linear operation is characteristic of the methods of multivariate analysis proper, which start with an interest in linear combinations, calculated for each unit, of the observed values of a number of variables.

Linear operations of this sort—on both units and variables—may also be involved in prediction of values of unobserved variables.

## §2. Vector space of variables

We will be analysing problems in which there is a fixed, finite set of variables, the same set for each unit. Supposing there are $p$ different variables, with $p > 1$, let $v_1, \ldots, v_p$ denote their *names*. Three examples of such sets are:

  (i)   a simple anthropometric set with $p = 2$,
        $v_1 = $ 'log weight in lbs'
        $v_2 = $ 'log height in inches';
  (ii)  a biochemical set with $p = 3$,
        $v_1 = $ 'initial velocity of reaction in mols per second'
        $v_2 = $ 'concentration of molecule M'
        $v_3 = $ 'square of concentration of molecule M';
  (iii) a geological set with $p = 5$,
        $v_1 = $ 'percentage of rock type A'
        $v_2 = $ 'percentage of rock type B'
        $v_3 = $ 'complementary percentage $100 - v_1 - v_2$'
        $v_4 = $ 'porosity measure $\alpha$'
        $v_5 = $ 'porosity measure $\beta$'.

Different units that happen to have equal values on each variable will be treated as equivalent and put in the same equivalence class of units. Such an equivalence class will be called an *evaluator* for which the symbol $e$ will be reserved. We will write

$$[e, v_1], \ldots, [e, v_p]$$

for the values corresponding to $v_1, \ldots, v_p$, respectively, and note that $e$ is specified by this vector in $R^p$.

  Given any $p$ real numbers $\lambda_1, \ldots, \lambda_p$, the composite expression $\lambda_1 v_1 + \ldots + \lambda_p v_p$ will be interpreted as the name of a generally new variable, defined simply by the values it has, for any evaluator. These values are taken to satisfy

$$[e, \lambda_1 v_1 + \ldots + \lambda_p v_p] = \lambda_1 [e, v_1] + \ldots + \lambda_p [e, v_p] \qquad (1)$$

for any evaluator $e$. Thus, for the anthropometric example (i), the new variable with name 'log weight $- 2.2$ log height' is assigned the value $[e, v_1] - 2.2[e, v_2]$ for evaluator $e$. The construction is consistent in the sense that, for example, $v_1$ and $1v_1 + 0v_2 + \ldots + 0v_p$ have the same values. It is, moreover, implicit in the construction based on eqn (1) that any newly constructed variable is either not observable directly or, if it were, its observed value would be consistent with the other observations $[e, v_1], \ldots, [e, v_p]$.

  The constructed value $[e, $ 'log weight $- 2.2$ log height'$]$ is typical of the derived quantities studied in multivariate analysis. The strength of the scientific case for their consideration is crucially dependent on the subject-matter context. Their applicability is, however, appreciably enhanced by the possibility that linear combinations of variables may represent a useful first-order approximation, in conditions of ignorance as to which non-linear functions would be required for exact scientific truthfulness.

The construction (1) gives us, by definition, a *vector space of variables* over the real field, of dimension $p$ and with $v_1, \ldots, v_p$ as a basis. We will always use $\mathcal{V}$ to denote this vector space. Our first small step to coordinate freedom will be to drop any reference to, or dependence on, the basis $v_1, \ldots, v_p$. As a result, our first picture, Fig. 2, is not particularly challenging.

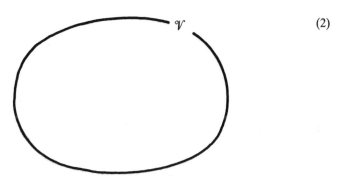

$$\mathcal{V} \tag{2}$$

## §3. Dual space of evaluators

Inspection of eqn (1) of §2 shows that the evaluator $e$ may be identified with the linear functional on $\mathcal{V}$ that takes values $[e, v_1], \ldots, [e, v_p]$ for the basis $v_1, \ldots, v_p$. It may therefore be positioned in the $p$-dimensional dual vector space, $\mathcal{V}'$, that consists of all the linear functionals on $\mathcal{V}$. This space will play such a central role in the development that it deserves better recognition than is afforded by the merely superscripted $\mathcal{V}'$. So we will use $\mathcal{E}$, instead of $\mathcal{V}'$, and refer to all the elements of $\mathcal{E}$ as 'evaluators', and to $\mathcal{E}$ itself as *evaluator space*. The symbol $e$ will then denote any one of these evaluators, and the bilinear form $[e, v]$ (Halmos 1958, p. 21) will still be called the *value* of $e$ on variable $v$. Given two evaluators $e_1$, $e_2$ and real $\lambda_1$, $\lambda_2$, the construct $\lambda_1 e_1 + \lambda_2 e_2$ will denote the evaluator defined by

$$[\lambda_1 e_1 + \lambda_2 e_2, v] = \lambda_1[e_1, v] + \lambda_2[e_2, v] \tag{1}$$

for any variable $v$.

It is fairly clear that $\mathcal{E}$ may be found to be housing generalized evaluators that violate some logical condition satisfied by the values of any unit-based evaluator $e$. For example, in the biochemical example (ii) of §2, a unit-based $e$ must satisfy $[e, v_3] = [e, v_2]^2$, while in the geological example, a necessary condition is that $[e, v_1] + [e, v_2] + [e, v_3] = 100$. Whether the violation of such conditions matters will depend on what we actually do with $\mathcal{V}$, $\mathcal{E}$ and the structures and operators that these vector spaces will be required to accommodate. Suffice to say that what is being touched on here is the well-known range of statistical absurdities such as 'the average family that has 2.1 children'.

*Exercises*

1. Tom, Dick, and Harry have had their weights and heights measured, with the results:

|  | Tom | Dick | Harry |
|---|---|---|---|
| Weight (lbs) | 154 | 210 | 123 |
| Height (ins) | 71 | 75 | 64 |

(i) Find real $\alpha$, $\beta$ such that Tom $= \alpha$ Dick $+ \beta$ Harry (in which equation the individuals' names denote the appropriate evaluators in the dual of the variable space for 'weight' and 'height').

(ii) How do $\alpha$ and $\beta$ change when the variables are taken instead to be 'log weight' and 'log height'?

(iii) Find and interpret the real $\lambda$ such that the variable

$$\text{lambdamix} =_{\text{def}} \lambda \log \text{weight} + (1 - \lambda) \log \text{height}$$

satisfies

$$[\text{Tom, lambdamix}] = [\tfrac{1}{2} \text{Dick} + \tfrac{1}{2} \text{Harry, lambdamix}]$$

2. Which parts (if any) of Exercise 1 survive when variable space is extended to include 'shoe size'?

## §4. Statistical problems

Even within the primitive framework so far developed, it is possible to state, at least in outline form, some simple statistical problems. At the core of these is the concept of a random evaluator $x \in \mathcal{E}$, distributed according to some probability distribution $P$ on $\mathcal{E}$. There is nothing esoteric in this concept: such an $x$ is in one–one correspondence with the real $p$-vector of values $[x, v_1], \ldots, [x, v_p]$, so that $P$ is nothing more than the coordinate-free entity determined by a probability distribution of this vector. More concretely, $P$ can correspond to choice of $x$ at random from some finite population of evaluators.

The mean or expectation of $x$ thus distributed may be defined as follows. Supposing

$$\int \|[x, v]\| \, dP < \infty$$

for all $v \in \mathcal{V}$, we have, for $u, v \in \mathcal{V}$ and any real $\alpha, \beta$

$$\int [x, \alpha u + \beta v] \, dP = \alpha \int [x, u] \, dP + \beta \int [x, v] \, dP$$

so that the function of $v$ given by $\int [x, v] \, dP$ is a linear functional on $\mathcal{V}$. It follows that there is a uniquely determined evaluator $\mu \in \mathcal{E}$ such that

$$[\mu, v] \equiv \int [x, v] \, dP. \tag{1}$$

We call $\mu$ the *mean* or *expectation* of $x$, and write $\mu = E(x) = \int x \, dP$.

Our first statistical problem is that of the *affine estimation* of $\mu$, from the information provided by a single observation $x$ distributed according to $P$. Specifically, we have to choose $a$ and $B$ in

$$\hat{\mu} = a + Bx,$$

where $a \in \mathcal{E}$ and $B$ is linear, $\mathcal{E} \to \mathcal{E}$. It is, supposedly, given that $\mu \in \mathcal{L}$, a proper (linear) subspace of $\mathcal{E}$. (The case in which it is given that $\mu$ lies in some affine manifold is covered by a change of origin in $\mathcal{E}$.) This problem becomes the coordinate-free version of the general linear model once the counterpart of the error variance–covariance matrix has been specified. In the present formulation, the 'error' will be the evaluator $f$ in the equation

$$x = \mu + f. \tag{2}$$

As it turns out, the coordinate-free encapsulation of the variance–covariance matrix may be given the simpler description of 'variance' without any sense of linguistic plagiarism. The additional information needed for one particular and important solution of the first problem is just the knowledge of this 'variance', at least up to proportionality.

Let us pause to record the elements of our first problem in Fig. 3. Note that, in (3), the vector space $\mathcal{L}$ has been portrayed as a line, serving adequately to display the vectors 0 and $\mu$. Expansion of $\mathcal{L}$ would give us the equivalent representation shown in (4).

To introduce our second, related problem, that of *affine prediction*, we will use a figure, (5), straight away. We have obtained (5) from (3) by the introduction of another subspace $\mathcal{H}$, for which we have portrayed the coset

$$x + \mathcal{H} =_{\text{def}} \{x + h : h \in \mathcal{H}\}.$$

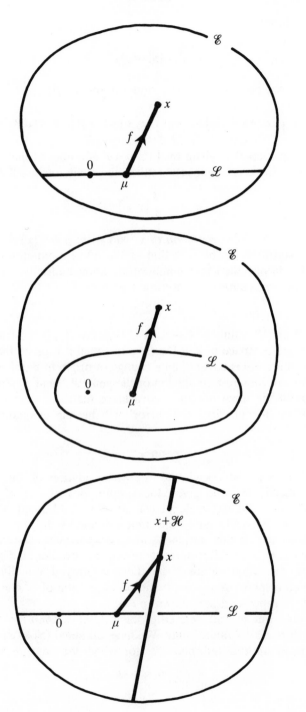

(3)

(4)

(5)

This coset may or may not intersect $\mathscr{L}$, and it is important that (5) should not mislead the viewer: the break in the line of $\mathscr{L}$, at the point where intersection would otherwise occur, serves to maintain neutrality about whether $(x + \mathscr{H}) \cap \mathscr{L} = \varnothing$ (the null set) or not. Figure 5 is also silent about whether $\mathscr{H} \cap \mathscr{L} = \{0\}$ or not.

In terms of (5), the prediction problem may be stated as follows. While we do know that $x$ is distributed according to $P$ and that the unknown mean $\mu$ lies in $\mathscr{L}$, we have only a partial observation of $x$, corresponding to knowledge of which of the cosets $\mathscr{E}/\mathscr{H}$ it lies in. An affine predictor of $x$, as a whole, i.e. as a completely specified vector in $\mathscr{E}$, would then be required to be of the form

$$\hat{x} = a + Bx$$

where $B(x + h) \equiv Bx$ for all $x \in \mathscr{E}$ and $h \in \mathscr{H}$, since $B$ must annihilate $\mathscr{H}$ in order that $\hat{x}$ be calculable from the given data.

Of course, for both the problem of affine estimation and that of affine prediction, we still have to develop the coordinate-free algebra for the assessment of the closeness of $\hat{\mu}$ and $\hat{x}$ to $\mu$ and $x$, respectively. These assessments are required for any optimal choice of $a$ and $B$.

Just as, in univariable statistics, ignorance about probability distributions is overcome by exploiting the information in a random sample, so here, when the 'variance' is not known (up to proportionality), it will be necessary to consider ways of dealing with a random sample of evaluators $x_1, \ldots, x_n$ from $P$.

It is clear that we have not yet even indicated the lines of any coordinate-free solutions of the problems just described. But it may also be apparent that it should be possible to uncover the solutions without recourse to the familiarities of their coordinatized analysis.

*Exercises*

1. A random vector $\mathbf{Y}$ in $R^n$ has expectation $\mathbf{X\theta}$ where $\mathbf{X}$ is a fixed $n \times k$ matrix and $\mathbf{\theta} \in R^k$ is unknown. Establish the coordinate-free representation of this, with $p = \dim \mathscr{V} = n$, by making $v_i$ the variable measured for the $i$th coordinate of $\mathbf{Y}$ so that $Y_i = [x, v_i]$, $i = 1, \ldots, n$. If $\mathbf{\alpha} \in \mathscr{A} \subset R^n \Leftrightarrow \mathbf{X'\alpha} = \mathbf{0}$, show that

$$\mathscr{L} = \bigcap_{\alpha \in \mathscr{A}} \mathscr{L}_\alpha$$

where

$$\mathscr{L}_\alpha = \left\{ e : [e, \sum_{i=1}^{n} \alpha_i v_i] = 0 \right\}.$$

2. Suppose that, in Exercise 1, $n = 3$, $k = 1$ and

$$\mathbf{X\theta} = \begin{bmatrix} \theta_1 \\ 2\theta_1 \\ 3\theta_1 \end{bmatrix}.$$

For the prediction problem in which only $Y_1$ is observed and prediction of $(Y_2, Y_3)$ is required, show that, in the coordinate-free rendering of the problem, $\mathcal{H} \cap \mathcal{L} = \{0\}$ and $(x + \mathcal{H}) \cap \mathcal{L} \neq \varnothing$ for all $x$. Draw an informative specialization of Fig. 5 for this case.

If $Y_1$, $Y_2$, $Y_3$ were independent normal with variance 1 and prediction of $Y_3$ from observation of $(Y_1, Y_2)$ were required, show that $(x + \mathcal{H}) \cap \mathcal{L} = \varnothing$ with probability 1.

3. Devise an example of the general model of Exercise 1 for which, in the coordinate-free rendering, $\mathcal{H} \cap \mathcal{L} \neq \{0\}$ and $(x + \mathcal{H}) \cap \mathcal{L} = \varnothing$ with probability 1. How do you resolve the difficulty of pictorial representation in this case?

4. Show that the condition $\mathcal{H} \cap \mathcal{L} = \{0\}$ is equivalent to estimability (in the usual sense of least-squares theory) of all the components of $\mathbf{X\beta}$ from the observations of the linear combinations of $\mathbf{Y}$ whose nullity defines $\mathcal{H}$. What is the equivalent of the condition $(x + \mathcal{H}) \cap \mathcal{L} = \varnothing$?

# 2
# Dualities

In this chapter, we set out, in geometrical fashion, some linear and bilinear algebra that relates operations in $\mathscr{E}$ to operations in $\mathscr{V}$. This apparatus will allow $\mathscr{V}$ to play an important role in the analysis of the statistical problems just described.

## §5. Values and operators

The following definitions involve the value function $[\ ,\ ]$ of §3 and linear transformations on and between the spaces $\mathscr{V}$ and $\mathscr{E}$.

**Definition 1**   Given a subspace $\mathscr{U}$ of $\mathscr{V}$, the subspace $\mathscr{U}^{\square}$ of $\mathscr{E}$ given by

$$\mathscr{U}^{\square} = \{e : [e, v] = 0 \quad \text{for all} \quad v \in \mathscr{U}\}$$

is the *bi-orthogonal complement* (or *annihilator*) of $\mathscr{U}$.

Likewise, we may define the bi-orthogonal complement in $\mathscr{V}$ of a subspace of $\mathscr{E}$. The dimension of $\mathscr{U}^{\square}$ is $p - \dim \mathscr{U}$. With the introduction of $\mathscr{U}^{\square}$, we now have the excuse to present a joint portrait, (1), of $\mathscr{V}$ and $\mathscr{E}$, with some related furniture in each space.

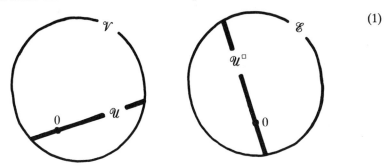

$$(1)$$

In (1), the related subspaces $\mathscr{U}$ in $\mathscr{V}$ and $\mathscr{U}^{\square}$ in $\mathscr{E}$ are represented by straight lines through the origin 0. We have made these representations orthogonal on the Euclidean page only as a remainder of the '-orthogonal' in the term 'bi-orthogonal'.

**Notation**   We will use $\mathscr{N}_A$ and $\mathscr{R}_A$ to denote the *null* and *range* space, respectively, of a linear transformation $A$.

9

The following two definitions are specializations of the general definition (Halmos 1958, §47, Exercise 1).

**Definition 2**   The *dual* (or *adjoint*) of the linear transformation $A$, $\mathcal{V} \to \mathcal{V}$ is the linear transformation $A'$, $\mathcal{E} \to \mathcal{E}$, characterized by

$$[A'e, v] = [e, Av]$$

for all $v \in \mathcal{V}$ and $e \in \mathcal{E}$. (Similarly for $A$, $\mathcal{E} \to \mathcal{E}$.)

**Definition 3**   The *dual* (or *adjoint*) of the linear transformation $B$, $\mathcal{V} \to \mathcal{E}$, is the linear transformation $B'$, $\mathcal{V} \to \mathcal{E}$, characterized by

$$[B'u, v] = [Bv, u]$$

for all $u$ and $v$ in $\mathcal{V}$. (Similarly for $B$, $\mathcal{E} \to \mathcal{V}$.)

The pictorial mnemonics are shown in Fig. 2. Note that $B'$ is $\mathcal{V} \to \mathcal{E}$ not $\mathcal{E} \to \mathcal{V}$, as ideas of symmetry might suggest. Conventionally, we will identify transformations by means of labelled arrows showing the direction of the action.

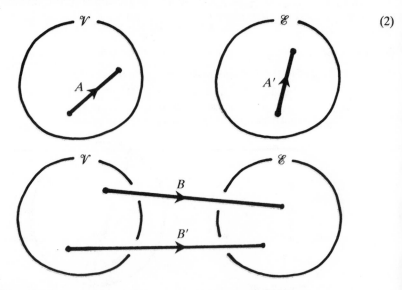

(2)

**Theorem 1**   For the transformations $A$ and $B$ of Definitions 2 and 3, $\mathcal{N}_A = \mathcal{R}_{A'}{}^{\square}$ and $\mathcal{N}_B = \mathcal{R}_{B'}{}^{\square}$.

**Proof**   $v \in \mathcal{N}_A \Leftrightarrow 0 \equiv [e, Av] \equiv [A'e, v] \Leftrightarrow v \in \mathcal{R}_{A'}{}^{\square}$.   A   similar   proof suffices for $B$.                                                                    $\square$

**Definition 4**   The linear transformation $B$, $\mathcal{V} \to \mathcal{E}$, is *symmetric* if $B = B'$.

Theorem 1 then implies that $\mathcal{N}_B = \mathcal{R}_B{}^\square$ when $B$ is symmetric. Note that a linear transformation $A$, $\mathcal{V} \to \mathcal{V}$ or $\mathcal{E} \to \mathcal{E}$, cannot be said to be symmetric.

**Definition 5** The linear transformation $B$, $\mathcal{V} \to \mathcal{E}$, is *positive* if $[Bv, v] > 0$ for all $v \neq 0$. It is *non-negative* if $[Bv, v] \geqslant 0$ for all $v$.

**Theorem 2** A symmetric, non-negative, linear transformation $B$, $\mathcal{V} \to \mathcal{E}$, is invertible if and only if it is positive.

**Proof** This follows from the equivalence, for symmetric non-negative $B$,

$$Bv = 0 \Leftrightarrow [Bv, v] = 0. \tag{3}$$

The $\Rightarrow$ in (3) is immediate. To establish the $\Leftarrow$, suppose $[Bv_0, v_0] = 0$. Then, by the non-negativity of $B$, we must have, at $\lambda = 0$, $0 = (d/d\lambda)[B(v_0 + \lambda v), v_0 + \lambda v] = 2[Bv_0, v]$ for all $v$. Hence $Bv_0 = 0$. □

An analogous definition and theorem may clearly be stated for $B$, $\mathcal{E} \to \mathcal{V}$.

We end this section with a definition of projection and the characterization of its dual. We start with projection in $\mathcal{E}$ and characterize its dual in $\mathcal{V}$, but the rôles of $\mathcal{E}$ and $\mathcal{V}$ are interchangeable.

**Definition 6** The linear transformation $\Pi$, $\mathcal{E} \to \mathcal{E}$, is *projection onto* $\mathcal{R}_\Pi$ *parallel to* $\mathcal{N}_\Pi$ if $\Pi e = e$ for all $e \in \mathcal{R}_\Pi$.

**Theorem 3** The dual $\Pi'$ of $\Pi$ is projection onto $\mathcal{N}_\Pi{}^\square$ parallel to $\mathcal{R}_\Pi{}^\square$.

The proof is straightforward, as is the associated Fig. 4.

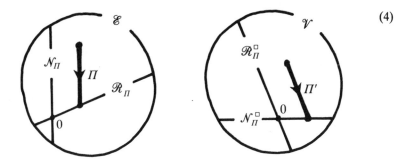

$$\tag{4}$$

*Exercises*

1. If $\mathcal{S}$ and $\mathcal{T}$ are any subspaces of $\mathcal{E}$ then
$$\mathcal{S}^\square + \mathcal{T}^\square = (\mathcal{S} \cap \mathcal{T})^\square.$$

2. If $\mathcal{E} = \mathcal{S} \oplus \mathcal{T}$ then $\mathcal{V} = \mathcal{S}^\square \oplus \mathcal{T}^\square$.

3. $B$, linear $\mathcal{V} \rightarrow \mathcal{E}$, is symmetric if and only if $[Bu, v] \equiv [Bv, u]$.

4. Defining the *rank* of a linear transformation, $\mathcal{E}$ or $\mathcal{V}$ to $\mathcal{E}$ or $\mathcal{V}$, as the dimension of its range, show that a linear transformation and its dual have the same rank.

5. Suppose $\mathcal{U}$ is a proper subspace of $\mathcal{V}$, where $\mathcal{V}$ is generated by variables $v_1, v_2, v_3$. Suppose $e_0 \in \mathcal{E}$ is defined by $[e_0, v_1] = 5$, $[e_0, v_2] = 7$, $[e_0, v_3] = 51$. How would you describe, in statistical terms, the set of evaluators $e_0 + \mathcal{U}^\square$?

6. Suppose $A$, linear $\mathcal{E} \rightarrow \mathcal{E}$, is idempotent, i.e. $A^2 = A$. Show that $A$ is a projection.

7. For $A$, linear $\mathcal{E} \rightarrow \mathcal{E}$ or $\mathcal{V}$, and $\mathcal{S}$ a subspace of $\mathcal{E}$, show that $A'(A\mathcal{S})^\square = \mathcal{S}^\square$ if and only if $\mathcal{N}_A \subset \mathcal{S}$.

## §6. Inner products

Suppose that $x \in \mathcal{E}$ is randomly distributed with probability distribution $P$ such that $E([x, v]^2) < \infty$ for all $v$ in $\mathcal{V}$. Define the function $V$, $\mathcal{V} \times \mathcal{V} \rightarrow R$, by

$$V(u, v) = \text{cov}([x, u], [x, v]).\tag{1}$$

It then follows that $V$ has the properties

$$V(u, v) \equiv V(v, u) \qquad \text{(symmetry)},\tag{2}$$

$$\begin{cases} V(\alpha u + \beta v, w) \equiv \alpha V(u, w) + \beta V(v, w) \\ V(w, \alpha u + \beta v) \equiv \alpha V(w, u) + \beta V(w, v) \end{cases} \quad \text{(bilinearity)},\tag{3}$$

$$V(v, v) \geq 0 \qquad \text{(non-negativity)}.\tag{4}$$

**Definition 1** $V$, defined by (1), will be called the *variance inner product generated by the probability distribution P,* or the *variance of P* for short. It may be written $\text{var}(x)$ when it is understood that $x$ has the distribution $P$.

**Definition 2** The variance inner product $V$ is *non-singular* when $V(v, v) > 0$ for $v \neq 0$; it is *singular* if $V(v, v) = 0$ for some $v \neq 0$.

We will find it convenient to allow the term 'inner product' to encompass the singular case, thereby deviating from the standard usage (Halmos 1958, §61). We will return to the variance inner product after some general definitions and associated theorems.

**Definition 3** The function $I$, $\mathcal{V} \times \mathcal{V} \rightarrow R$, is an *inner product on* $\mathcal{V}$ if it is symmetric, bilinear and non-negative. It is *non-singular* or *singular* as for $V$ in Definition 2.

**Definition 4**  For a given inner product $I$ on $\mathscr{V}$, the *associated linear transformation, I,* is the linear transformation, $\mathscr{V} \to \mathscr{E}$, characterized by

$$[Iu, v] = I(u, v)$$

for all $u$, $v$.

There is little risk of confusion but significant notational advantage in using the same symbol $I$ for the two interpretations—'inner product' and 'associated linear transformation'. The two roles of $I$ are brought together by the following readily established connections:

(i)  As a transformation, $\mathscr{V} \to \mathscr{E}$, an inner product is non-negative, symmetric, while any non-negative, symmetric transformation, $\mathscr{V} \to \mathscr{E}$, provides an inner product on $\mathscr{V}$.

(ii)  Non-singularity and singularity of an inner product correspond to positivity and its negation, respectively, of the associated linear transformation, which, by Theorem 2 of §5, correspond to the latter's invertibility and non-invertibility, respectively.

The next theorem uses both interpretations of $I$.

**Theorem 1**  $\mathscr{N}_I =_{\text{def}} \{v : Iv = 0\} = \{v : I(v, v) = 0\}$.

**Proof**  $I$ is symmetric and non-negative, whence the proof of Theorem 2 of §5 shows that $Iv = 0 \Leftrightarrow I(v, v) = 0$.  □

**Definition 5**  Given an inner product $I$ on $\mathscr{V}$, the *I-orthogonal complement, $\mathscr{U}^I$,* of a subspace $\mathscr{U}$ of $\mathscr{V}$ is the subspace

$$\mathscr{U}^I = \{v : I(v, u) = 0 \text{ for all } u \in \mathscr{U}\}.$$

When $I$ is non-singular, $\mathscr{U}$ and $\mathscr{U}^I$ are complementary, and there is a simple picture, (5), that records their relationship. In Fig. 5, we have

(5)

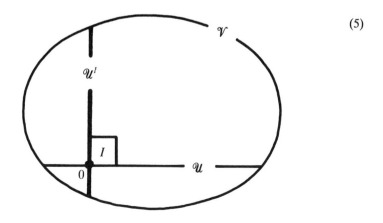

used the 'buttress' symbol $\boxed{I}$, which serves here and elsewhere to denote the $I$-orthogonality of any two flats.

When $I$ is singular, $\mathcal{N}_I \subset \mathcal{U}^I$ and $\mathcal{N}_I \cap \mathcal{U}$ may not be null: how then to devise a pictorial representation is left as an exercise for the reader.

The next result further illustrates the interplay of the two roles of $I$.

**Theorem 2**   $I\mathcal{U}^I \subseteq \mathcal{U}^\square$ with equality when $I$ is non-singular.

**Proof**   For $v \in \mathcal{U}^I$, we have $I(v, u) = [Iv, u] = 0$ for all $u \in \mathcal{U}$, whence $I\mathcal{U}^I \subseteq \mathcal{U}^\square$. When $I$ is invertible and $e \in \mathcal{U}^\square$, we have $[II^{-1}e, u] = 0$ for all $u = \mathcal{U}$, whence $I^{-1}e \in \mathcal{U}^I$ and $\mathcal{U}^\square \subseteq I\mathcal{U}^I$.   $\square$

A related result will also be useful.

**Theorem 3**   $(I\mathcal{U}) \cap \mathcal{U}^\square = \{0\}$.

**Proof**   When $I$ is invertible and the associated inner product is non-singular, $\mathcal{U} \cap \mathcal{U}^I = \{0\}$ and a simple argument suffices:

$$\{0\} = I(\mathcal{U} \cap \mathcal{U}^I) = (I\mathcal{U}) \cap (I\mathcal{U}^I) = (I\mathcal{U}) \cap \mathcal{U}^\square$$

by Theorem 2. However, an indirect proof is required when $I$ is non-invertible: if $(I\mathcal{U}) \cap \mathcal{U}^\square \neq \{0\}$, there is $u \in \mathcal{U}$, with $Iu \neq 0$, such that $[Iu, u] = 0$, which contradicts Theorem 1.   $\square$

The following result concerns any two $I$-orthogonal subspaces:

**Theorem 4**   If $\mathcal{U}$, $\mathcal{W}$ are $I$-orthogonal subspaces of $\mathcal{V}$ (that is $I(u, w) \equiv 0$ for $u \in \mathcal{U}$, $w \in \mathcal{W}$) then

$$I\mathcal{U} \subseteq \mathcal{W}^\square \qquad \text{and} \qquad I\mathcal{W} \subseteq \mathcal{U}^\square.$$

**Proof**   $I(u, w) \equiv [Iu, w] \equiv 0$ for $u \in \mathcal{U}$, $w \in \mathcal{W} \Rightarrow Iu \in \mathcal{W}^\square$. Similarly $Iw \in \mathcal{U}^\square$.   $\square$

*Exercises*

1. If $I_1$ and $I_2$ are any two inner products on $\mathcal{V}$, show that $\mathcal{N}_{I_1+I_2} = \mathcal{N}_{I_1} \cap \mathcal{N}_{I_2}$ and $\mathcal{R}_{I_1+I_2} = \mathcal{R}_{I_1} + \mathcal{R}_{I_2}$.

2. For $I$ a singular inner product on $\mathcal{V}$, devise a general pictorial representation of $\mathcal{U}$, $\mathcal{U}^I$ and $\mathcal{N}_I$ in $\mathcal{V}$. (*Suggestion*: Use 'three dimensions' to accommodate $\mathcal{U}$, $\mathcal{N}_I$ are their intersection, and a dotted line through $0$ as a complement of $\mathcal{N}_I$ in $\mathcal{U}^I$, 'dotted' to indicate that it lives in 'higher dimensions' than those of $\mathcal{U} + \mathcal{N}_I$.)

3. For $I$ an inner product on $\mathcal{V}$, and $A$, linear $\mathscr{E} \to \mathscr{E}$, show that $AIA'$ is the associated linear transformation of another inner product on $\mathcal{V}$. What if $A$ is, instead, $\mathscr{E} \to \mathcal{V}$?

4. Given a random vector $x \in \mathscr{E}$ with non-singular variance $V$, and a

proper subspace $\mathscr{U}$ of $\mathscr{V}$, show that the $V$-orthogonal projection of $v$ on $\mathscr{U}$ has maximal 'correlation with $v$' of any variable in $\mathscr{U}$, where the correlation of $u$ and $v$ means the correlation coefficient of their randomly distributed values $[x, u]$, $[x, v]$.

## §7. Dual inner product and orthogonal projection

The linear transformation, $\mathscr{V} \to \mathscr{E}$, associated with a non-singular inner product $I$ on $\mathscr{V}$ has an inverse $I^{-1}$ which may be shown to be symmetric and positive.

**Definition 1**   Given a non-singular inner product $I$ on $\mathscr{V}$, the *dual inner product* is the inner product on $\mathscr{E}$ associated with the linear transformation $I^{-1}$. (The term 'dual inner product' is adopted reluctantly, since it is not the inner product corresponding to the *adjoint* dual of $I$ as linear transformation.)

It may be verified that $I(u, v) \equiv I^{-1}(Iu, Iv)$, an identity that can be described as saying that $I$ (as transformation) carried $I$ (as inner product) into $I^{-1}$ (as inner product).

**Theorem 1**   If $I$ is a non-singular inner product on $\mathscr{V}$ and the subspaces $\mathscr{U}$ and $\mathscr{W}$ of $\mathscr{V}$ are $I$-orthogonal then the subspaces $I\mathscr{U}$ and $I\mathscr{W}$ of $\mathscr{E}$ are $I^{-1}$-orthogonal. Moreover, $\mathscr{U}$ and $I\mathscr{W}$ are bi-orthogonal, as are $\mathscr{W}$ and $I\mathscr{U}$.

The proof is as straightforward as the picture, (1).

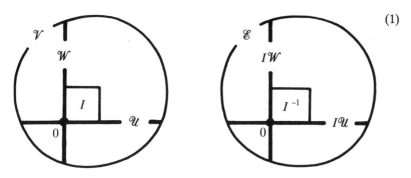

(1)

**Definition 2**   Given a non-singular inner product $I$ on $\mathscr{V}$ and a subspace $\mathscr{U}$ of $\mathscr{V}$, *I-orthogonal projection onto* $\mathscr{U}$ is the projection onto $\mathscr{U}$ parallel to $\mathscr{U}^I$.

**Theorem 2**   For $I$ non-singular, $I$-orthogonal projection onto $\mathscr{U}$ is the linear transformation $\Pi$, $\mathscr{V} \to \mathscr{V}$ and $\mathscr{R}_\Pi = \mathscr{U}$, uniquely determined by either

    (i)  $\Pi$ minimizes $I(v - \Pi v, v - \Pi v)$ for all $v \in \mathcal{V}$, or
    (ii)  $I(u, v - \Pi v) = 0$ for all $u \in \mathcal{U}, v \in \mathcal{V}$.

The proof is straightforward.

The proof of the next theorem uses the fact that, for $I$ non-singular, the null space, $\mathcal{U}^I$, of $I$-orthogonal projection onto $\mathcal{U}$ is equal to $(I\mathcal{U})^\square$. This identity may be derived from Theorem 2 of §6 with $\mathcal{U}^I$ in place of $\mathcal{U}$.

**Theorem 3**  For non-singular inner product $I$ on $\mathcal{V}$, the dual of $I$-orthogonal projection onto $\mathcal{U}$ is $I^{-1}$-orthogonal projection onto $I\mathcal{U}$.

**Proof**  By Theorem 3 of §5, the dual of projection onto $\mathcal{U}$ with null space $(I\mathcal{U})^\square$ is projection onto $I\mathcal{U}$ with null space $\mathcal{U}^\square$. By Theorem 2, of §6, $\mathcal{U}^\square = I\mathcal{U}^I$ and, by Theorem 1 here, the latter is $I^{-1}$-orthogonal to $I\mathcal{U}$.    $\square$

There is a type of symmetry associated with $I$-orthogonal projection that, because projection does not 'change spaces', cannot be that of Definition 4 of §5. The required definition, stated in a form that allows wider application, is

**Definition 3**  A linear transformation $A$, $\mathcal{V} \to \mathcal{V}$, is *I-symmetric* with respect to an inner product $I$ on $\mathcal{V}$ if $IA = A'I$.

We may note that if $A$ is $I$-symmetric with respect to a *non-singular* inner product $I$, then $A'$ is $I^{-1}$-symmetric. It is instructive to contrast $I$-symmetry with the '$I$-isometry' of

**Definition 4**  A linear transformation $A$, $\mathcal{V} \to \mathcal{V}$, is *I-isometric* with respect to an inner product $I$ on $\mathcal{V}$ if $I(Au, Av) = I(u, v)$ for all $u, v$.

**Theorem 4**  $A$ is $I$-isometric if and only if $A'IA = I$.

**Theorem 5**  If $A$ is $I$-isometric for non-singular $I$ then $A'$ is $I^{-1}$-isometric.

The pictures for $I$-symmetry and $I$-isometry on the left and right of (2), respectively, make an interesting contrast. The $I$-symmetry of our $I$-orthogonal projection, $\Pi$, is readily established, and leads to the joint portrayal of $\Pi$ and $\Pi'$ in Fig. 3. What can be said about $I$-orthogonal projection when $I$ is singular? The answer can be summarized as follows. When $I$ is singular, $\mathcal{U}$ and $\mathcal{U}^I$ have a non-trivial intersection that is a subset of $\mathcal{N}_I$, and 'projection onto $\mathcal{U}$ parallel to $\mathcal{U}^I$' is not uniquely defined. However, for $v \in \mathcal{V}$, the set of such projections of $v$ on $\mathcal{U}$ is a translate of $\mathcal{U} \cap \mathcal{U}^I$ and is equal to the set of values of $u \in \mathcal{U}$ that minimize $I(v - u, v - u)$. Further analysis is deferred to Exercise 6. The

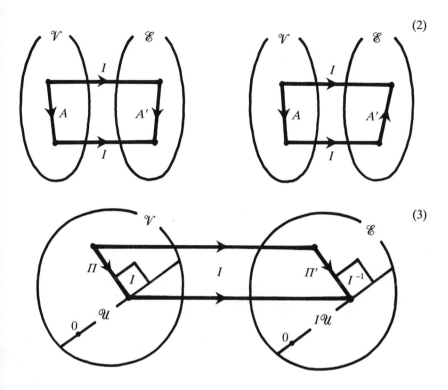

question of the dual of any such $I$-orthogonal projection will be left until we have found a generalization of $I^{-1}$ for singular $I$.

*Exercises*

1. The linear transformation $A$, $\mathcal{V} \to \mathcal{V}$, is $I$-symmetric if and only if

$$I(Au, v) \equiv I(Av, u).$$

2. If $I$ and $J$ are any inner products on $\mathcal{V}$ and $\mathcal{E}$, respectively, then $JI$, linear $\mathcal{V} \to \mathcal{V}$, is $I$-symmetric.

3. If $A$, linear $\mathcal{V} \to \mathcal{V}$, is both $I$-symmetric and $I$-isometric with respect to a non-singular inner product $I$ on $\mathcal{V}$ then $A$ is self-inverse (that is, $A^2 = 1$, the identity transformation).

4. If for $A$, linear $\mathcal{E} \to \mathcal{E}$, and all non-singular inner products $I_1$, $I_2$ on $\mathcal{V}$, we have

$$I_1^{-1}AI_1 = I_2^{-1}AI_2$$

then $A = \alpha 1$, a scalar multiple of the identity transformation. (Hint: choose $e_0 \in \mathcal{E}$ and take $I_2$ defined by $I_2 v \equiv [e_0, v]e_0 + I_1 v$.)

(4)

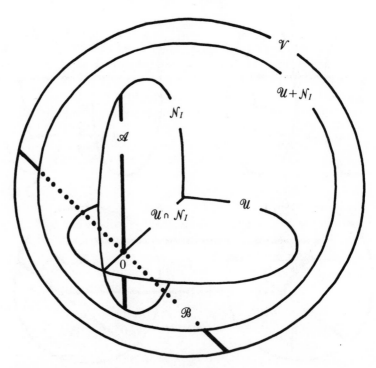

5. In justification of Fig. 3, prove that

$$\Pi' = I\Pi I^{-1}.$$

6. *I-orthogonal projection on $\mathscr{U}$ for singular $I$.* The relevant picture, (4), is that suggested for Exercise 2 of §6. Here $\mathscr{B}$ is the partly dotted line representing any complement of $\mathscr{N}_I$ in $\mathscr{U}^I$ so that $\mathscr{U}^I = \mathscr{B} \oplus \mathscr{N}_I$, and $\mathscr{A}$ is any subspace of $\mathscr{N}_I$ complementary to $\mathscr{U} \cap \mathscr{N}_I$. Define $(I, \mathscr{A}, \mathscr{B})$-*orthogonal projection onto $\mathscr{U}$* to be projection onto $\mathscr{U}$ parallel to $\mathscr{A} \oplus \mathscr{B}$. Show that, if we define an *I-orthogonal projection of $v \in \mathscr{V}$ on $\mathscr{U}$* as any $u_0 \in \mathscr{U}$ that minimizes $I(v - u, v - u)$, the set $\{u_0\}$ is a coset in $\mathscr{U}$ of $\mathscr{U} \cap \mathscr{N}_I$, and show that this coset is the set of $(I, \mathscr{A}, \mathscr{B})$-orthogonal projections of $v$ on $\mathscr{U}$.

7. If $I$ is an inner product on $\mathscr{V}$, then $I(u, v)^2 \leqslant I(u, u)I(v, v)$. Use this inequality to prove that, if $I$ is non-singular,

$$|[e, v]| \leqslant \{I^{-1}(e, e)\, I(v, v)\}^{\frac{1}{2}}$$

for any $e \in \mathscr{E}$, $v \in \mathscr{V}$.

8. If $I$ is a non-singular inner product on $\mathscr{V}$, and $\mathscr{S}$ is any subspace of $\mathscr{V}$, then

$$(\mathscr{S}^I)^{\square} = (\mathscr{S}^{\square})^{I^{-1}}.$$

## §8. Dualtors and their integral representations

In §6, we saw how inner products such as $V$ and $I$ play two roles—either as bilinear forms on $\mathcal{V} \times \mathcal{V}$ or as linear transformations, $\mathcal{V} \to \mathcal{E}$. Such dual role operators, or *dualtors* as they may be dubbed, will now be given a concise integral representation that will turn out to be particularly felicitous for statistical purposes. The representation has much in common with that for tensor product discussed by Halmos (1958, §25). However, the tensor connection will not be developed, for the simple reason that we do not need anything like the full power of the associated pure mathematical machinery.

A linear combination of vectors in evaluator space $\mathcal{E}$ corresponds to the same linear combinations of their values on any choosen variable in $\mathcal{V}$. In the same spirit, but in reverse, we may ask what operations on vectors in $\mathcal{E}$ correspond to the routine statistical operations of taking squares and cross-products of their values. Starting with a single evaluator $x$ in $\mathcal{E}$, these operations yield quantities such as $[x, v]^2$ and $[x, u][x, v]$. These are all obtainable from the bilinear form, written $x^2$ and defined by

$$x^2(u, v) \equiv [x, u][x, v].$$

Likewise, for two evaluators $x$, $y$, we may define $xy$, $\mathcal{V} \times \mathcal{V} \to R$, by

$$xy(u, v) \equiv [x, u][y, v].$$

Such bilinear forms may then be extended to a vector space over the real field. For $x_1, y_1, \ldots, x_n, y_n$ in $\mathcal{E}$ and $\alpha_1, \ldots, \alpha_n$ real, we define

$$\alpha_1 x_1 y_1 + \ldots + \alpha_n x_n y_n, \quad \mathcal{V} \times \mathcal{V} \to R,$$

by

$$(\alpha_1 x_1 y_1 + \ldots + \alpha_n x_n y_n)(u, v) \equiv \alpha_1[x_1, u][y_1, v] + \ldots + \alpha_n[x_n, u][y_n, v].$$

The most general element, $A$ say, of this space of bilinear operators may be written

$$A = \int_\Omega x(\omega)y(\omega)\,d\alpha(\omega) \tag{1}$$

where $\alpha$ is a signed measure on some measurable space $\Omega$ and $x(.)$ and $y(.)$ are measurable functions. In its first role, the action of $A$ is then

$$A(u, v) \equiv \int_\Omega [x(\omega), u][y(\omega), v]\,d\alpha(\omega), \tag{2}$$

the integrals existing if

$$\int_{\Omega} [x(\omega), v]^2 |d\alpha(\omega)| < \infty \quad \text{and} \quad \int_{\Omega} [y(\omega), v]^2 |d\alpha(\omega)| < \infty$$

for all $v \in \mathscr{V}$.

It is possible to give any bilinear form on $\mathscr{V} \times \mathscr{V}$ an integral representation of the form (1), in an infinity of ways.

The second role of $A$, that of linear transformation, $\mathscr{V} \to \mathscr{E}$, is now defined by

$$Av \equiv \int_{\Omega} x(\omega)[y(\omega), v] d\alpha(\omega) \tag{3}$$

(integrals existing). Any linear transformation, $\mathscr{V} \to \mathscr{E}$, may be represented by (3), in form and action. Moreover, the dual $A'$, linear $\mathscr{V} \to \mathscr{E}$ also, then has the representation

$$A' = \int_{\Omega} y(\omega) x(\omega) d\alpha(\omega). \tag{4}$$

Note that, in (3), we have adopted the definition $xy(v) = x[y, v]$ for $xy$ as a linear operator.

It is clear that we now have the promised integral representation of dualtors: it remains to be seen how useful it is. The extension of the above definitions to accommodate transformations $\mathscr{E} \times \mathscr{E} \to R$, $\mathscr{E} \times \mathscr{V} \to R$, $\mathscr{V} \times \mathscr{E} \to R$, $\mathscr{E} \to \mathscr{V}$, $\mathscr{E} \to \mathscr{E}$, $\mathscr{V} \to \mathscr{V}$ is immediate.

A more radical extension of (1)–(4) is represented by having $x$ in $\mathscr{E}$ but $y$ in a general finite-dimensional vector space $\mathscr{G}$. In that case, $A$ is both

(i) a bilinear form on $\mathscr{V} \times \mathscr{G}'$

$$A(v, g') = \int_{\Omega} [x(\omega), v][y(\omega), g'] d\alpha(\omega)$$

(ii) a linear transformation, $\mathscr{G}' \to \mathscr{E}$,

$$Ag' = \int x(\omega)[y(\omega), g'] d\alpha(\omega)$$

(integrals existing).

*Exercises*

1. Given $A = \int_{\Omega} x(\omega) y(\omega) d\alpha(\omega)$, $\mathscr{V} \to \mathscr{E}$, and $B = \int_{\bar{\Omega}} u(\bar{\omega}) v(\bar{\omega}) d\beta(\bar{\omega})$, $\mathscr{E} \to \mathscr{V}$, what is the integral representation of $AB$?

2. If $v_1, \ldots, v_p$ span $\mathscr{V}$ and $e_1, \ldots, e_p$ in $\mathscr{E}$ satisfy $[e_i, v_j] = \delta_{ij}$, show

that $\sum_{i=1}^{p} e_i v_i = 1$, the identity transformation on $\mathscr{E}$. For $A = \sum_{i=1}^{p} e_i^2$, $B = \sum_{i=1}^{p} v_i^2$, show that $A = B^{-1}$.

## §9. Inner products as integrals

Our first application of the integral representation (1) of §8 is to the variance inner product for a probability distribution $P$ on $\mathscr{E}$.

Identifying $\Omega$ with $\mathscr{E} = \{x\}$, setting $d\alpha(\omega) = dP$ and taking $x(\omega) = y(\omega) = x - \mu$ where $\mu = \int x \, dP$, the representation (1) of §8 becomes $\int (x - \mu)^2 \, dP$. Then, by definition,

$$\left( \int (x - \mu)^2 \, dP \right)(u, v) \equiv \int [x - \mu, u][x - \mu, v] \, dP$$

$$= \mathrm{cov}([x, u], [x, v])$$

$$= V(u, v).$$

So, we can confidently make the identification

$$V = \int (x - \mu)^2 \, dP. \tag{2}$$

This formulation of $V = \mathrm{var}(x)$ will be heavily exploited.

Can a similar formulation be achieved for the covariance of two jointly distributed random vectors $x$ and $y$ on $\mathscr{E}$? Retaining $\Omega$ for the underlying probability space, and writing $x = x(\omega)$, $y = y(\omega)$, $\mu = \int x(\omega) \, dP(\omega)$, $v = \int y(\omega) \, dP(\omega)$, $\alpha = P$, the representation becomes

$$\int (x(\omega) - \mu)(y(\omega) - v) \, dP(\omega) = C,$$

say. We have

$$C(u, v) \equiv \mathrm{cov}([x, u], [y, v]).$$

The dualtor $C$ will be called the 'covariance of $x$ and $y$' and written $\mathrm{cov}(x, y)$ when we need to indicate the ordered random vectors it is defined for. Note that, in general, $C$ will not equal its dual $C'$, i.e. $\mathrm{cov}(x, y) \neq \mathrm{cov}(y, x)$. The conditions $\mathrm{cov}(x, y) = 0$ and $\mathrm{cov}(y, x) = 0$ are equivalent and, when they obtain, $x$ and $y$ are said to be *uncorrelated*.

Another important dualtor that benefits from such integral, or sum, representations is the *sample variance*, $S$, defined by

$$S = \frac{1}{(n-1)} \sum_{i=1}^{n} (x_i - \bar{x})^2, \tag{2}$$

from a random sample $x_1, \ldots, x_n$ on $\mathscr{E}$ with $\bar{x} = (x_1 + \ldots + x_n)/n$.

To obtain an integral representation of the dualtor $V^{-1}$, in the case when $V$ is non-singular/invertible, we first use $V^{-1}$ to transfer $P$ from $\mathscr{E}$ to $\mathscr{V}$.

**Definition 1**  When $V$ is invertible, the *dual error distribution* $Q$ on $\mathscr{V}$, corresponding to $P$ on $\mathscr{E}$, is defined by

$$Q(.) = P(\mu + V(.))\qquad(3)$$

where the argument in $(.)$ is any measurable subset of $\mathscr{V}$. (Invertibility is necessary in order that $Q$, thus defined, be a probability distribution.)

It is then verifiable that

$$V^{-1} = \int v^2 \, \mathrm{d}Q.\qquad(4)$$

For the dualtor provided by a general inner product $I$ on $\mathscr{V}$, there is an infinity of integral representations

$$I = \int e^2 \, \mathrm{d}\alpha\qquad(5)$$

where $\alpha$ is a non-negative measure on $\mathscr{E}$. Since $\alpha$ is thus not a 'given', as $P$ is for $V$, we have no interest in defining any analogue of $Q$. In place of (4), we have the equally tautological

**Theorem 1**  If $I = \int e^2 \, \mathrm{d}\alpha$ is non-singular, then $I^{-1} = \int (I^{-1}e)^2 \, \mathrm{d}\alpha$.

**Proof**  For $e_1, e_2 \in \mathscr{E}$,

$$\left(\int (I^{-1}e)^2 \, \mathrm{d}\alpha\right)(e_1, e_2) = \int [e_1, I^{-1}e][e_2, I^{-1}e] \, \mathrm{d}\alpha$$

$$= \int [e, I^{-1}e_1][e, I^{-1}e_2] \, \mathrm{d}\alpha$$

$$= I(I^{-1}e_1, I^{-1}e_2) = I^{-1}(e_1, e_2).$$

(Alternatively, apply the general formula of Exercise 1.)          □

An important application of the integral representation concerns a decomposition of $I$, and an associated necessary and sufficient condition for $I^{-1}$-orthogonality of subspaces. If $\mathscr{S}$ and $\mathscr{T}$ are complementary subspaces of $\mathscr{E}$, let $s = s(e)$ and $t = t(e)$ denote the $\mathscr{S}$ and $\mathscr{T}$ components, respectively, of $e = s(e) + t(e)$. Then

$$e^2 = s^2 + st + ts + t^2$$

and

$$I = \int s^2 \, d\alpha + \int (st + ts) \, d\alpha + \int t^2 \, d\alpha.$$

Associated with this decomposition of $I$, we have

**Theorem 2** If $I = \int e^2 \, d\alpha$ is non-singular, a necessary and sufficient condition that the complementary subspaces $\mathscr{S}$ and $\mathscr{T}$ be $I^{-1}$-orthogonal is that

$$\int st \, d\alpha = 0 \Leftrightarrow \int ts \, d\alpha = 0 \Leftrightarrow \int (st + ts) \, d\alpha = 0, \tag{6}$$

equivalent to

$$I = \int s^2 \, d\alpha + \int t^2 \, d\alpha. \tag{7}$$

**Proof** Using $s^0$, $t^0$ to denote *variables*, in $\mathscr{S}^\square$, $\mathscr{T}^\square$ respectively,

$$I(t^0, s^0) = \int [e, t^0][e, s^0] \, d\alpha = \int [s, t^0][t, s^0] \, d\alpha$$

$$= \int [s, u][t, v] \, d\alpha = \left( \int st \, d\alpha \right)(u, v)$$

where $v$ is any variable with $\mathscr{S}^\square$-component equal to $s^0$ and $u$ is any variable with $\mathscr{T}^\square$-component equal to $t^0$. It follows that $\int st \, d\alpha = 0$ is equivalent to $I$-orthogonality of $\mathscr{S}^\square$, $\mathscr{T}^\square$, which is equivalent to $I^{-1}$-orthogonality of $\mathscr{S}$, $\mathscr{T}$. Trivially, $\int st \, d\alpha = 0 \Leftrightarrow \int ts \, d\alpha = 0 \Rightarrow \int (st + ts) \, d\alpha = 0$. Less so, $A =_{\text{def}} \int st \, d\alpha$ is linear into $\mathscr{S}$ and $B =_{\text{def}} \int ts \, d\alpha$ is linear into $\mathscr{T}$, whence $\mathscr{S} \cap \mathscr{T} = \{0\}$ means that $A + B = 0 \Rightarrow A = B = 0$. □

The general statement for any number of component spaces, up to $p$, is as follows. Its proof is left as an exercise.

**Theorem 3** If $I = \int e^2 \, d\alpha$ is non-singular and $\mathscr{E} = \mathscr{S}_1 \oplus \mathscr{S}_2 \oplus \ldots \oplus \mathscr{S}_s$, $2 \leqslant s \leqslant p$, a necessary and sufficient condition that $\mathscr{S}_1, \ldots, \mathscr{S}_s$ be pairwise $I^{-1}$-orthogonal is

$$\int s_i s_j \, d\alpha = 0 \qquad i \neq j, \, i = 1, \ldots, s, \, j = 1, \ldots, s,$$

where $s_i = s_i(e)$ is the component of $e$ in $\mathscr{S}_i$, $i = 1, \ldots, s$.

**Corollary** If $\mathscr{G}$ is a subspace of $\mathscr{E}$ and $\mathscr{G} = \mathscr{S}_1 \oplus \mathscr{S}_2$, a necessary and sufficient condition that $\mathcal{S}_1$ and $\mathcal{S}_2$ be $I^{-1}$-orthogonal is that $\int s_1 s_2 \, d\alpha = \int s_2 s_1 \, d\alpha = 0$, where $s_1, s_2$ are the components in $\mathcal{S}_1, \mathcal{S}_2$, respectively, of the $I^{-1}$-orthogonal projection of $e$ on $\mathcal{G}$.

Another application of the integral representation is in the definition of a real-valued evaluation between two inner products, $I$ on $\mathcal{V}$ and $J$ on $\mathcal{E}$.

**Definition 2**  Given inner products $I$ on $\mathcal{V}$ and $J$ on $\mathcal{E}$ with $I = \int e^2 \, d\alpha$, $J = \int v^2 \, d\beta$, we define

$$[I, J] = \iint [e, v]^2 \, d\alpha \, d\beta. \tag{8}$$

The double integral in (8) may be written in either of the alternative forms $\int I(v, v) \, d\beta$, $\int J(e, e) \, d\alpha$, showing that it is, as the notation of (8) suggests, a function of $I$ and $J$ independent of their representations. The non-negative quantity $[I, J]$ is, in fact, the evaluation of the tensor $I$ at the bilinear form $J$, when $I$ is regarded as a linear functional on the vector space of bilinear forms on $\mathcal{E} \times \mathcal{E}$ (Halmos 1958, §25).

**Theorem 4**  If $I$ is a non-singular inner product on $\mathcal{V}$,

$$[I, I^{-1}] = \dim \mathcal{V}. \tag{9}$$

**Proof**  We have $\dim \mathcal{V} = p$. By Exercise 4, we may write $I = e_1^2 + \ldots + e_p^2$, where $e_1, \ldots, e_p$ are $I^{-1}$-orthonormal, whence, by Theorem 1, $I^{-1} = (I^{-1}e_1)^2 + \ldots + (I^{-1}e_p)^2$ and

$$[I, I^{-1}] = \sum \sum [e_i, I^{-1}e_j]^2 = p. \tag{10}$$

$\square$

Finally, we state without proof the equivalent of the standard canonical reduction of a non-negative-definite quadratic form.

**Theorem 5**  If $I$ is a non-singular and $H$ a general inner-product on $\mathcal{V}$, then $\{e_1, \ldots, e_p\}$ may be found so that

$$H = \alpha_1 e_1^2 + \ldots + \alpha_p e_p^2,$$
$$I = e_1^2 + \ldots + e_p^2,$$

where $HI^{-1}e_i = \alpha_i e_i$ and $\alpha_i = H(I^{-1}e_i, I^{-1}e_i)$, $i = 1, \ldots, p$.

*Exercises*

1. If $I = \int e^2 \, d\alpha$ and $A, B$ are linear, $\mathcal{E} \to \mathcal{E}$ or $\mathcal{E} \to \mathcal{V}$, then

$$\int (Ae)(Be) \, d\alpha = AIB'.$$

2. If $I = e_1^2 + \ldots + e_n^2$ is non-singular,

$$\sum_{i=1}^{n} I^{-1}(e_i, e)e_i = e \qquad \text{for all } e \in \mathscr{E} \qquad (11)$$

$$\sum_{i=1}^{n} I^{-1}(e_i, e)I^{-1}(e_i, f) = I^{-1}(e, f) \qquad \text{for all } e \in \mathscr{E}, f \in \mathscr{E}. \qquad (12)$$

3. Prove Theorem 3.

4. Let $\mathscr{E} = \mathscr{S}_1 \oplus \ldots \oplus \mathscr{S}_p$ where $\mathscr{S}_1, \ldots, \mathscr{S}_p$ are 1-dimensional, $I^{-1}$-orthogonal subspaces. For $e_i \in \mathscr{S}_i$ with $I^{-1}(e_i, e_i) = 1$, $i = 1, \ldots, n$, show that

$$I = e_1^2 + \ldots + e_p^2.$$

5. If $I$ is non-singular, show that $[I, J]$ is the trace of the matrix of the transformation $JI$, $\mathscr{V} \to \mathscr{V}$, with respect to any $I$-orthonormal basis of $\mathscr{V}$.

6. Show that $[e^2, J] = J(e, e)$.

7. If $x$ and $y$ are two jointly distributed random evaluators in $\mathscr{E}$ then

$$\text{var}(x + y) = \text{var}(x) + \text{var}(y) + \text{cov}(x, y) + \text{cov}(y, x).$$

8. If $x$ is a random evaluator in $\mathscr{E}$ with non-singular variance $V$ and $x_1, \ldots, x_s$ are its respective components for the direct sum $\mathscr{E} = \mathscr{S}_1 \oplus \ldots \oplus \mathscr{S}_s$, where $\mathscr{S}_1, \ldots, \mathscr{S}_s$ are $V^{-1}$-orthogonal, then

$$\text{var}(x) = \text{var}(x_1) + \ldots + \text{var}(x_s).$$

9. Continuing Exercise 7, find an example with $p = 2$ in which $\text{cov}(x, y) + \text{cov}(y, x) = 0$ but $\text{cov}(x, y) \neq 0$. (*Hint*: It is sufficient to take three points $\omega_1$, $\omega_2$, $\omega_3$ in the underlying probability space for $(x(\omega), y(\omega))$, with an equiprobability distribution on $\omega_1$, $\omega_2$, $\omega_3$.)

10. Suppose $x \in \mathscr{E}$, distributed with mean $\mu$ and non-singular variance $V$, lies with probability 1 on the $V^{-1}$-sphere

$$V^{-1}(x - \mu, x - \mu) = \alpha.$$

Show that $\alpha = p$, the dimension of $\mathscr{E}$.

11. Show that, when $\mu \neq 0$ and $V$ is non-singular, $V^{-1}\mu$ is a variable whose value has the smallest coefficient of variation, if we exclude variables whose values have zero expectation.

## §10. Shadow, marginal, and transformed inner products

Suppose we are given an inner product $I$ on $\mathscr{V}$, with integral representation $I = \int e^2 \, d\alpha$. We may use Exercise 1 of §9 to generate some important related inner products, each associated with some linear transformation of $\mathscr{E}$.

Thus, for the transformation $A$, $\mathscr{E} \to \mathscr{E}$, the derived dualtor

$$I^A =_{\text{def}} \int (Ae)^2 \, d\alpha \qquad (1)$$

is also an inner product on $\mathscr{V}$. The associated linear transformation is given by $I^A = AIA'$. This identity shows that the terminology $I^A$ is appropriate in that (1) is a function of $I$ and $A$, the same for all representations of $I$. We will call $I^A$ the *shadow inner product* derived from $I$ by transformation $A$. This follows Dempster (1969, p. 118), although our definition differs slightly from his. The term 'shadow' is somewhat misleading in its suggestion of projection—we allow $A$ to be *any* linear transformation, $\mathscr{E} \to \mathscr{E}$.

Closely related to the concept of a shadow inner product is another type of derived inner product, obtained by 'marginalization' of $\mathscr{E}$ as follows. Given a proper subspace $\mathscr{T}$ of $\mathscr{E}$, it is convenient to abuse notation and write

$$\tau[e] = e + \mathscr{T} \qquad (2)$$

to denote both (i) the translate of $\mathscr{T}$ that contains $e$ and (ii) the linear transformation $e \to \tau[e]$ for which $\tau[e]$ is then regarded as the element of the quotient vector space $\mathscr{E}/\mathscr{T}$ corresponding to the coset $e + \mathscr{T}$. With the latter interpretation in mind, we define the $\mathscr{T}$-*marginal inner product* by

$$I^{\mathscr{T}} = \int (\tau[e])^2 \, d\alpha. \qquad (3)$$

For $\tau \in \mathscr{E}/\mathscr{T}$, the entity $\tau^2$ is a dualtor acting on $(\mathscr{E}/\mathscr{T})'$ or $(\mathscr{E}/\mathscr{T})' \times (\mathscr{E}/\mathscr{T})'$, just as $e^2$ acts on $\mathscr{V}$ or $\mathscr{V} \times \mathscr{V}$. Along lines suggested at the end of Exercise 4(b) of §22 of Halmos (1958), it can be analytically convenient to identify, isomorphically, $(\mathscr{E}/\mathscr{T})'$ and $\mathscr{T}^{\square}$. This is done by defining

$$[\tau[e], w] = [e, w] \qquad (4)$$

for $e \in \mathscr{E}$, $w \in \mathscr{T}^{\square}$. For vectors in $\mathscr{T}^{\square}$, the inner products $I$ and $I^{\mathscr{T}}$ then coincide.

Both shadow and marginal inner products are specializations of what we will call *transformed inner products*. The general form of these may be defined by

$$I^B = \int (Be)^2 \, d\alpha, \qquad (5)$$

where $B$ is linear from $\mathscr{E}$ to some finite-dimensional vector space $\mathscr{G}$. In (5), $I^B$ is an inner product on $\mathscr{G}'$, the dual of $\mathscr{G}$. Its associated linear transformation, $\mathscr{G}' \to \mathscr{G}$, is $BIB'$ where $B'$ is the dual (or adjoint) of $B$

defined by

$$[e, B'g'] = [Be, g']$$

for all $e \in \mathscr{E}$ and $g' \in \mathscr{G}'$ (Halmos 1958, §44).

We complete this section with two theorems involving shadow inner products. The first is simple, the second more technical. Both theorems will find application in later chapters.

**Theorem 1**  If $B$ is $I^{-1}$-orthogonal projection onto a $q$-dimensional subspace $\mathscr{S} \subset \mathscr{E}$,

$$[I^B, I^{-1}] = q.$$

**Proof**  We may write

$$I = s_1^2 + \ldots + s_q^2 + t_1^2 + \ldots + t_{p-q}^2$$

where $s_1, \ldots, s_1, t_1, \ldots, t_{p-q}$ are $I^{-1}$-orthonormal and $\{s_i\}$ are in $\mathscr{S}$. Then $Bs_i = s_i$, $i = 1, \ldots, q$, and $Bt_j = 0$, $j = 1, \ldots, p - q$, so that $I^B = s_1^2 + \ldots + s_q^2$, while

$$I^{-1} = (I^{-1}s_1)^2 + \ldots + (I^{-1}s_q)^2 + (I^{-1}t_1)^2 + \ldots + (I^{-1}t_{p-q})^2.$$

Hence

$$[I^B, I^{-1}] = \sum_{i=1}^{q} \sum_{j=1}^{q} [s_i, I^{-1}s_j]^2 + \sum_{i=1}^{q} \sum_{j=1}^{p-q} [s_i, I^{-1}t_j]^2$$

$$= q. \qquad \qquad \square$$

There is a natural topology in $\mathscr{E}$ and in $\mathscr{V}$, provided by any coordinatization. The following result implicitly uses this topology, in describing the limiting properties of the dual of a particular 'augmented' singular inner product as the 'augmentation' making it non-singular is reduced to zero. Along with the main result, which introduces an important shadow inner product, the statement of the theorem incorporates some technical results that will be of use later.

**Theorem 2**  Given two inner products on $\mathscr{V}$, namely, singular $S$ and non-singular $I$, define an *'augmented dual'*, $A(\rho)$, of $S$ by

$$A(\rho) = (S + \rho I)^{-1},$$

$\rho > 0$. Then, with $J =_{\mathrm{def}} I^{-1}$,

(i)  $\rho A(\rho)$ inverts $I$ between $\mathscr{N}_S$ and $\mathscr{R}_S^J$,

(ii)  $A(\rho)\mathscr{R}_S = \mathscr{N}_S^I$,

(iii)  restricted to action on $\mathscr{R}_S$, $\lim_{\rho \to 0} A(\rho)$ exists and inverts $S$ between $\mathscr{N}_S^I$ and $\mathscr{R}_S$,

(6)

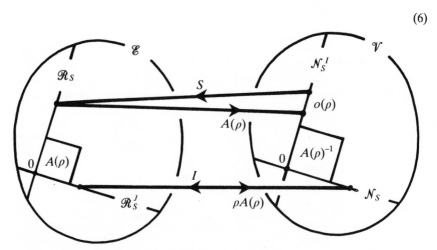

(iv) $\mathcal{N}_S$ and $\mathcal{N}_S^I$ are also $A(\rho)^{-1}$-orthogonal,

(v) $\mathcal{R}_S$ and $\mathcal{R}_S^J$ are also $A(\rho)$-orthogonal,

(vi) $\lim_{\rho \to 0} \rho A(\rho)$ exists and equals the shadow inner product $J^{\Pi}$ where $\Pi$ is $I$-orthogonal projection onto $\mathcal{N}_S$.

Before proof, let us picture these claims in Fig. 6.

**Proof**

(i)   For $v \in \mathcal{N}_S$, $\rho A(\rho) I v = A(\rho)(A(\rho)^{-1} - S)v = v$.

(ii)  For $w \in \mathcal{N}_S^I$, $A(\rho)^{-1} w = Sw + \rho I w$ which is in $\mathcal{R}_S$ because both $Sw$ and $Iw$ are. The result then follows from the fact that $\dim \mathcal{R}_S = \dim \mathcal{N}_S^I$.

(iii) holds because, *restricted to action between* $\mathcal{N}_S^I$ *and* $\mathcal{R}_S$, $S + \rho I$ is non-singular for $\rho \geq 0$ and continuous in $\rho$.

(iv)  For $v \in \mathcal{N}_S$ and $w \in \mathcal{N}_S^I$

$$A(\rho)^{-1}(v, w) = \rho[(\rho A(\rho))^{-1} v, w] = \rho I(v, w) = 0.$$

(v)   That $\mathcal{R}_S = A(\rho)^{-1} \mathcal{N}_S^I$ and $\mathcal{R}_S^J = A(\rho)^{-1} \mathcal{N}_S$ then implies their $A(\rho)$-orthogonality, by Theorem 1 of §7.

(vi)  For fixed $e$ in $\mathcal{R}_S$, $v =_{\text{def}} \rho A(\rho)e$ in $\mathcal{N}_S^I$ satisfies $\rho^{-1} Sv + Iv = e$, whence $v \to 0$ as $\rho \to 0$ (Exercise 7). Hence $\rho A(\rho)e \to 0 = \Pi J \Pi' e$, trivially. For $e$ in $\mathcal{R}_S^J$, $\rho A(\rho)e = I^{-1}e = \Pi J \Pi' e$, trivially. Hence, for all $e$, $\rho A(\rho)e \to J^{\Pi} e$.   $\square$

*Exercises*

1. If $I$ is a non-singular inner product on $\mathcal{V}$ and if $A$, linear $\mathcal{E} \to \mathcal{E}$, is such that $\mathcal{R}_A \cap \mathcal{R}_{1-A} = \{0\}$, show that $I = I^A + I^{1-A}$ if and only if $A$ is $I^{-1}$-orthogonal projection.

2. For its action on vectors in a subspace $\mathscr{F}$ of $\mathscr{E}$, the dualtor $I^{-1}$ is determined by knowledge of $(\mathscr{F}^\square)^I$ and $I^{\Pi'}$, where $\Pi'$ is $I^{-1}$-orthogonal projection onto $\mathscr{F}$ (see Fig. 3 of §7).

3. Make sense of the equations $(I^A)^B = I^{BA}$, $(I^{\mathscr{S}})^{\mathscr{T}} = I^{\mathscr{T}}$, and justify the identities $(I^{-1})^I = I$, $(I)^{I^{-1}} = I^{-1}$.

4. If $B$ is linear, $\mathscr{E} \to \mathscr{G}$, identify $C$ such that

$$I^B = (I^{\mathscr{N}_B})^C.$$

5. Investigate the possibilities of

$$I^{B_1 + B_2} = I^{B_1} + I^{B_2},$$

generalizing Exercise 1, for the two cases

(i) $B_1$ and $B_2$ both linear, $\mathscr{E} \to \mathscr{G}$,
(ii) $B_1$ linear, $\mathscr{E} \to \mathscr{G}_1$, and $B_2$ linear, $\mathscr{E} \to \mathscr{G}_2$, and $(B_1 + B_2)e$ $=_{\mathrm{def}} B_1 e + B_2 e \in \mathscr{G}_1 \oplus \mathscr{G}_2$.

6. If $H$ is a singular inner product on $\mathscr{V}$ and $J$ is any non-singular inner product on $\mathscr{E}$, show that

$$\mathscr{N}_{H'} = (\mathscr{R}_H)^J.$$

7. Establish the assertion '$v \to 0$ as $\rho \to 0$' in the proof of Theorem 2 $(vi)$.


## §11. Singularities

We will now study singular variances and inner products, with no more justification for the moment than is provided by a forward reference to some statistical examples in Exercise 7, and the intuition that such study is likely to be useful when we eventually tackle the decoordinatization of generalized inverses. Such inverses find application in many parametrized problems even when, as is usually the case, the variance inner product is non-singular.

The singularity status of a variance inner product is determined by the size of the manifold spanned by the support of the underlying probability distribution.

**Theorem 1**    The variance inner product

$$V = \int (x - \mu)^2 \, \mathrm{d}P$$

is singular or not, according as the support of $P$ is or is not confined to a proper affine manifold of $\mathscr{E}$.

**Proof**   $V(v, v) = 0 \Leftrightarrow \int [x - \mu, v]^2 \, dP = 0$

$$\Leftrightarrow P([x - \mu, v] = 0) = 1$$
$$\Leftrightarrow \text{the support of } x - \mu \text{ lies in } \langle v \rangle^\square,$$

where $\langle v \rangle$ is the one-dimensional subspace of $\mathscr{V}$ spanned by $v$.   □

The analogous result for the sample variance $S$ may be stated less indirectly: $S$ is non-singular or singular, according to whether the vectors $x_1 - \bar{x}, \ldots, x_n - \bar{x}$ span $\mathscr{E}$ or only a proper subspace of $\mathscr{E}$, respectively.

Taking the adequately general case provided by singular $V$, our objective will be to uncover natural coordinate-free extensions of ingredients of the theory so far developed that have invoked non-singularity. In particular, we will look at the possible extensions of the concepts 'dual probability distribution' and 'dual inner product'. If our objective is attainable for $V$, it should then be a straightforward matter to apply the findings to $S$ or, to the extent that they are applicable, to any singular inner product.

Suppose then that the support of $P$ is confined to some translate of $\mathscr{F}$, where $\mathscr{F}$ is a proper subspace of $\mathscr{E}$. In other words, with probability 1, the associated random error $f = x - \mu$ lies in $\mathscr{F}$. We will refer to $\mathscr{F}$ as the *error subspace*, and suppose also that $\mathscr{F}$ is minimal in the sense that the support of $x - \mu$ spans $\mathscr{F}$.

We can start by relating $\mathscr{F}$ to the role of $V$ as associated linear transformation.

**Theorem 2**   $\mathscr{F} = \mathscr{N}_V{}^\square = \mathscr{R}_V$.

**Proof**   $v \in \mathscr{N}_V \Leftrightarrow \int (f, v)^2 \, dP = 0$

$$\Leftrightarrow [f, v] = 0 \quad \text{for } f \text{ in the support of } x - \mu$$
$$\Leftrightarrow [f, v] = 0 \quad \text{for } f \in \mathscr{F}. \qquad \square$$

In §9, we used the identity

$$Q(.) = P(\mu + V(.)) \tag{1}$$

to define the dual error distribution for the case where $V$ is invertible. Now, the non-invertibility of $V$ means that (1) will not give a probability distribution on $\mathscr{V}$, unless the class of subsets used as argument in $Q$ is restricted. A conducive restriction, and one that we will adopt, is to the subsets of $\mathscr{W}$, some subspace of $\mathscr{V}$ complementary to $\mathscr{N}_V$. The probability distribution on $\mathscr{W}$ that is then uniquely defined by (1) must, however, be referred to as *a* (rather than *the*) *dual error distribution*, in view of the arbitrariness in the choice of $\mathscr{W}$.

Let us define *a dual inner product* $V^+$ of $V$ by

$$V^+ = \int_W w^2 \, dQ \tag{2}$$

in analogy with eqn (4) of §9. The following theorem, while useful for our immediate objective, also lays some of the groundwork for the chapter on generalized inverses.

**Theorem 3** As a linear transformation, $\mathscr{E} \to \mathscr{W} \subset \mathscr{V}$, $V^+$ satisfies

(i) $V^+$ is the inverse of $V$ when both $V^+$ and $V$ are restricted to act between $\mathscr{F}$ and $\mathscr{W}$,

(ii) $\mathscr{N}_{V^+} = \mathscr{W}^\square$, that is, $V^+$ annihilates the subspace $\mathscr{W}^\square$, complementary to $\mathscr{F}$.

**Proof**

(i) Let $V^-$ denote the inverse of $V$ restricted to $\mathscr{W}$ and $\mathscr{F}$. For $w_0 \in \mathscr{W}$ and $f_0 = Vw_0 \in \mathscr{F}$,

$$V^+ f_0 = \int w[w, f_0] \, dQ(w) = \int_{\mu + \mathscr{F}} V^-(x - \mu)[V^-(x - \mu), Vw_0] \, dP(x)$$

$$= V^- \int_{\mu + \mathscr{F}} (x - \mu)[VV^-(x - \mu), w_0] \, dP(x)$$

$$= V^- \left( \int (x - \mu)^2 \, dP(x) \right) w_0$$

$$= V^- V w_0 = w_0$$

(thus $V^+$ agrees with $V^-$ in their action on $\mathscr{F}$).

(ii) From the definition, $V^+$ is symmetric, whence

$$\mathscr{N}_{V^+} = \mathscr{R}_{V^+}{}^\square = \mathscr{W}^\square. \qquad \square$$

The picture for Theorem 3 is given in Fig. 3. The very same construction can be carried out for a general singular inner product $I = \int e^2 \, d\alpha$, in which the support of the non-negative measure $\alpha$ spans a proper subspace, $\mathscr{R}_I$, of $\mathscr{E}$. We define *a dual inner product* $I^+$ on $\mathscr{E}$ by $I^+ = \int w^2 \, d\beta$, where $\beta$, the 'dual measure' on $\mathscr{W}$, is of only technical interest, unlike its counterpart $Q$ in the variance case. As a linear transformation, $\mathscr{E} \to \mathscr{V}$, $I^+$ has range $\mathscr{W}$, null space $\mathscr{W}^\square$ and is the inverse of $I$ when restricted to $\mathscr{W}$ and $\mathscr{R}_I$, which properties serve as a characterization. Theorem 1 of §9 carries over to give the self-referring identity

$$I^+ = \int (I^+ e)^2 \, d\alpha.$$

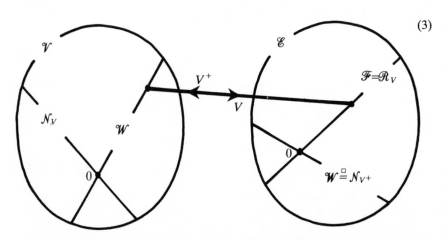

(3)

The non-uniqueness of $V^+$ and $I^+$ stems from the arbitrariness in the choice of $\mathcal{W}$. This affects $V^+$ and $I^+$ as linear transformations, and also as inner products, on $\mathcal{E}$ as a whole. However, it does not affect the inner products as applied to the important subspaces $\mathcal{F}(=\mathcal{R}_V)$ and $\mathcal{R}_I$, respectively, as the following theorem, stated in terms of $I$, shows.

**Theorem 4**  Restricted to vectors in $\mathcal{R}_I$, the inner product $I^+$ is independent of the choice of $I^+$ and is non-singular.

**Proof**  From the construction of $I^+$, we see that, for $e \in \mathcal{R}_I$, different choices of $I^+$ lead to differences in $I^+e$ that lie in $\mathcal{N}_I$. Hence the associated differences in $I^+(e, e) = [e, I^+e]$ are of the form $[e, v]$ where $v$ is in $\mathcal{N}_I$, which value is zero. Also, for non-zero $e \in \mathcal{R}_I$, we have $e = Iw$ with $w \neq 0$, and, by Theorem 1 of §6,

$$0 < [Iw, w] = [e, I^+e] = I^+(e, e).\qquad\qquad\square$$

The arbitrariness in the treatment of the singular case almost vanishes when we think of $I^+$ as the 'intrinsic' dual of $I$ when the evaluator space is reduced from $\mathcal{E}$ to $\mathcal{E}^* = \mathcal{R}_I$. Taking, for dual of $\mathcal{E}^*$, any subspace $\mathcal{V}^*$ of $\mathcal{V}$ complementary to $\mathcal{N}_I$, we have that $I$, as linear transformation $\mathcal{V}^* \to \mathcal{E}^*$, takes $I$ as inner product into $I^+$, whatever the choice of $\mathcal{V}^*$ (see the remark after Definition 1 of §7).

The viewpoint thus arrived at accommodates a useful 'singular' version of the Corollary of §9:

**Theorem 5**  If $\mathcal{G}$ is a subspace of $\mathcal{R}_I$ and $\mathcal{G} = \mathcal{S}_1 \oplus \mathcal{S}_2$, a necessary and sufficient condition that $\mathcal{S}_1$ and $\mathcal{S}_1$ be $I^+$-orthogonal is that $\int s_1 s_2\, d\alpha = \int s_2 s_1\, d\alpha = 0$, where $s_1$, $s_2$ are the components in $\mathcal{S}_1$, $\mathcal{S}_2$, respectively, of the $I^+$-orthogonal projection of $e \in \mathcal{R}_I$ on $\mathcal{G}$.

Under the conditions of this theorem, the decomposition previously noted for the non-singular case ((7) of §9) also holds, when $\mathcal{G} = \mathcal{R}_I$:

$$I = \int s_1^2 \, d\alpha + \int s_2^2 \, d\alpha. \tag{4}$$

Some useful algebraic properties of $I^+$, as linear transformation $\mathcal{E} \to \mathcal{V}$, follow directly from the construction:

$$II^+I = I, \tag{5}$$

$$I^+II^+ = I^+, \tag{6}$$

$I^+I$ is projection onto $\mathcal{W}$ parallel to $\mathcal{N}_I$, $\qquad$ (7)

$II^+$ is projection onto $\mathcal{R}_I$ parallel to $\mathcal{W}^\square$. $\qquad$ (8)

When $I$ is non-singular then, by Theorem 2 of §6, $I\mathcal{S}^\square = \mathcal{S}^{I^{-1}}$ where $\mathcal{S}$ is any subspace of $\mathcal{E}$. The analogue of this result, when $I$ is singular, is:

**Theorem 6** If $\mathcal{S}$ is any subspace of $\mathcal{R}_I \subset \mathcal{E}$ then $I\mathcal{S}^\square = \mathcal{S}^+$, where $\mathcal{S}^+$ is the $I^+$-orthogonal complement in $\mathcal{R}_I$ of $\mathcal{S}$.

**Proof** For $v \in \mathcal{S}^\square$ and $s \in \mathcal{S}$,

$$I^+(Iv, s) = [II^+s, v] = [s, v] = 0,$$

whence $I\mathcal{S}^\square \subset \mathcal{S}^+$. For $e \in \mathcal{S}^+$ and $s \in \mathcal{S}$,

$$0 = I^+(s, e) = [s, I^+e],$$

whence $I^+e \in \mathcal{S}^\square$ and $e = II^+e \in I\mathcal{S}^\square$. Hence $\mathcal{S}^+ \subset I\mathcal{S}^\square$ also. $\qquad\square$

Duals of singular inner products can be employed to give an explicit formula for orthogonal projection within $\mathcal{R}_I$ or, more generally, within any affine translate of $\mathcal{R}_I$. Suppose $I = \int e^2 \, d\alpha$. Let $\mathcal{T}$ be a subspace of $\mathcal{R}_I$, and $\mathcal{S}$ a subspace of $\mathcal{E}$ complementary to $\mathcal{T}$. The following theorem covers both cases, $I$ non-singular and singular.

**Theorem 7** the $I^{-1}$- or $I^+$-orthogonal projection of $e = s + t$ on $\mathcal{S} \cap (e + \mathcal{R}_I)$ is

$$s_0 = s - \left( \int st \, d\alpha \right) \left( \int t^2 \, d\alpha \right)^+ t \tag{9}$$

where $(\int t^2 \, d\alpha)^+$ is any dual inner product of $\int t^2 \, d\alpha$.

**Proof** Define $t_0 = t + (\int st \, d\alpha)(\int t^2 \, d\alpha)^+ t$. Then $e = s_0 + t_0$ and $s_0$, $t_0$ are components of $e$ with respect to $\mathcal{S}$ and the transformed complement $\mathcal{T}_0 = \{t_0 : t \in \mathcal{T}\} \subset \mathcal{R}_I$. We will show that $\mathcal{T}_0$ and $\mathcal{S} \cap \mathcal{R}_I$ are $I^{-1}$ or $I^+$-orthogonal. By Theorem 2 of §9 for the non-singular case or Theorem 5 of this section, with $\mathcal{G} = \mathcal{R}_I$, for the singular case, it will be sufficient to prove that $\int s_0 t_0 \, d\alpha = 0$. this integral is

$$\int st \, d\alpha - \left(\int st \, d\alpha\right)\left(\int t^2 \, d\alpha\right)^+ \int t^2 \, d\alpha + \int s \cdot \left(\int st \, d\alpha\right)\left(\int t^2 \, d\alpha\right)^+ t \, d\alpha$$
$$- \int \left\{\left(\int st \, d\alpha\right)\left(\int t^2 \, d\alpha\right)^+ t\right\}^2 d\alpha. \quad (10)$$

The third term in (10) is

$$\left(\int st \, d\alpha\right)\left(\int t^2 \, d\alpha\right)^+ \int ts \, d\alpha,$$

using

$$\int s(At) \, d\alpha = \left(\int st \, d\alpha\right)A'$$

for $A$ linear. Using

$$\int (At)^2 \, d\alpha = A\left(\int t^2 \, d\alpha\right)A'$$

and property (6), the fourth term is

$$- \left(\int st \, d\alpha\right)\left(\int t^2 \, d\alpha\right)^+ \left(\int t^2 \, d\alpha\right)\left(\int t^2 \, d\alpha\right)^+ \left(\int ts \, d\alpha\right)$$
$$= -\left(\int st \, d\alpha\right)\left(\int t^2 \, d\alpha\right)^+ \left(\int ts \, d\alpha\right),$$

cancelling the third. The second term is

$$- \int s \cdot \left(\int t^2 \, d\alpha\right)\left(\int t^2 \, d\alpha\right)^+ t \, d\alpha = - \int st \, d\alpha,$$

which cancels the first.

We use Fig. 11 to complete the proof. (In the non-singular case, $\mathcal{R}_I$ and $e + \mathcal{R}_I$ coincide.) The required projection is $e - t_0$ which equals $s_0$. □

Less straightforward is the characterization of $I^+$-orthogonal projection (see §7) onto $\mathcal{L}$, a subspace of $\mathcal{E}$ but not of $\mathcal{R}_I$. This problem is explored in Exercises 4 and 5.

Another useful representation of orthogonal projection can be found when the flat, onto which projection is made with respect to one inner product, is itself characterized as the range space of a second inner product. We will consider only $I^{-1}$-orthogonal projection onto a proper subspace $\mathcal{L}$ of $\mathcal{E}$, leaving the singular case for Exercise 6 and beyond. Suppose $\mathcal{L} = \mathcal{R}_U$ where $U$ is a singular inner product on $\mathcal{V}$.

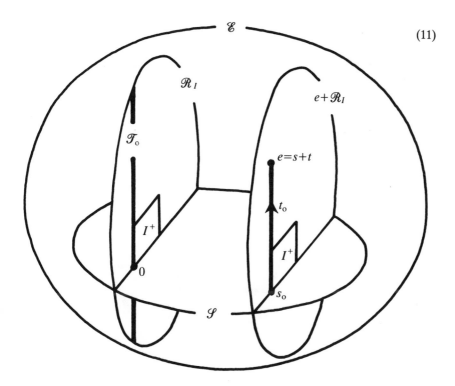

(11)

**Theorem 8** The $I^{-1}$-orthogonal projection of $e$ onto $\mathcal{L}$ is $\Pi e$ where

$$\Pi = U(UI^{-1}U)^{+}UI^{-1} \tag{12}$$

and $(UI^{-1}U)^{+}$ is any dual of the singular inner product $UI^{-1}U$ on $\mathcal{V}$.

**Proof** The proof is most easily obtained by verification of Fig. 13, in which two copies of $\mathcal{V}$ and of $\mathcal{E}$ are employed, for clarity of exposition. (Using only one copy of each is found to give a 'cat's cradle' that is hard to decipher.) Looking firstly at $\mathcal{E}$ at top right, we see the arbitrary $e$ and its $I^{-1}$-orthogonal projection $\Pi e$ on $\mathcal{R}_U$. The dualtor inner product $I^{-1}$ then transforms these vectors into two vectors, $d$ and $c$, in $\mathcal{V}$, whose difference lies in $\mathcal{N}_U$, because

$$e - \Pi e \in \mathcal{R}_U^{I^{-1}}$$

and

$$I^{-1}\mathcal{R}_U^{I^{-1}} = \mathcal{R}_U^{\square} = \mathcal{N}_U.$$

Thus the convergence on $b$ in the further transformation $U, \mathcal{V} \to \mathcal{E}$, is verified.

(13)

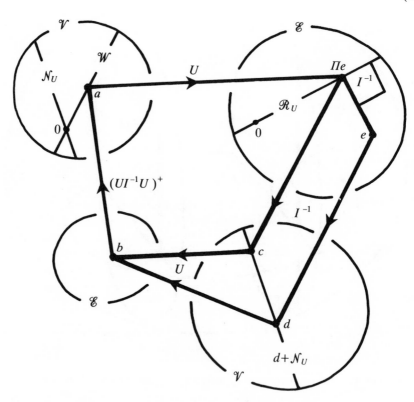

Looking next at $\mathcal{V}$ at top left, $\mathcal{W}$ is an arbitrary complement of $\mathcal{N}_U$ and $a$ is defined so that $Ua = \Pi e$.

We are now in a position to verify the labelling of the transformation from $b$ to $a$ as $(UI^{-1}U)^+$, since the transformation taking us from $a$ to $b$, via $\Pi e$ and $c$, is $UI^{-1}U$, and $\mathcal{R}_{UI^{-1}U} = \mathcal{R}_U$.

With $(UI^{-1}U)^+$ verified, we return to $e$ and see that $\Pi e = U(UI^{-1}U)^+UI^{-1}$, by going to $\Pi e$ the long way round.      □

The generalization of Theorem 4 of §9 is

**Theorem 9**   For any dual $I^+$ of $I$,

$$[I, I^+] =_{\dim} \mathcal{R}_I.$$

**Proof**   If $\dim \mathcal{R}_I = r$, we may write

$$I = e_1^2 + \ldots + e_r^2$$

where $e_1, \ldots, e_r \in \mathcal{R}_I$ are $I^+$-orthonormal. then

$$I^+ = (I^+ e_1)^2 + \ldots + (I^+ e_r)^2$$

and $[I, I^+] = \sum \sum [e_i, I^+ e_j]^2 = r$. □

The generalization of Theorem 1 of §10 is:

**Theorem 10** If $B$ is $I^+$-orthogonal projection onto a $q$-dimensional subspace $\mathcal{S} \subset \mathcal{R}_I$

$$[I^B, I^+] = q.$$

**Proof** With the notation of the proof of Theorem 9, we may suppose that $e_1, \ldots, e_q$ just span $\mathcal{S}$. Then, as in the proof of Theorem 1 of §10, $I^B = e_1^2 + \ldots + e_q^2$ and the result follows. □

The final item in this collection of 'singular' results is one that, like Theorem 7, brings together the concepts of projection and singular inner product, in this case a shadow (essentially marginal) inner product. We will have to wait until our final section to see an important application of its simple but technical content.

**Theorem 11** Given a non-singular inner product $I$ on $\mathcal{V}$, and two subspaces $\mathcal{R}$ and $\mathcal{S}$ of $\mathcal{E}$ with $\mathcal{E} = \mathcal{R} + \mathcal{S}$, let $\mathcal{C} = \mathcal{R} \cap \mathcal{S}$ and let $\mathcal{A}$ be any subspace of $\mathcal{S}$ complementary in $\mathcal{S}$ to $\mathcal{C}$. Let $A$ denote projection onto $\mathcal{R}$ parallel to $\mathcal{A}$, let $\Pi$ denote $I^{-1}$-orthogonal projection, $\mathcal{E} \rightarrow \mathcal{S}$, and let $\Pi^*$ denote $(I^A)^+$-orthogonal projection, $\mathcal{R} \rightarrow \mathcal{C}$. Then

$$A\Pi = \Pi^* A. \qquad (14)$$

**Proof** Let $\mathcal{T}$ be any subspace of $\mathcal{R}$ such that $\mathcal{E} = \mathcal{S} \oplus \mathcal{T} = \mathcal{A} \oplus \mathcal{C} \oplus T$. Then $e = s + t = a + c + t$ are the associated resolutions of $e$, while $Aa = 0$, $Ac = c$ and $At = t$. By Theorem 7, if $I = \int e^2 \, d\alpha$,

$$\Pi e = a + c - \left( \int (a+c) t \, d\alpha \right) \left( \int t^2 \, d\alpha \right)^+ t,$$

whence

$$A\Pi e = c - \left( \int ct \, d\alpha \right) \left( \int t^2 \, d\alpha \right)^+ t,$$

which, by another application of Theorem 7, is seen to be $\Pi^*(c + t)$ or $\Pi^* A e$. □

The content of Theorem 11 is shown in Fig. 15, in which $\square$, $\boxed{*}$ denote $I^{-1}$ and $(I^A)^+$ right-angles, respectively.

(15)

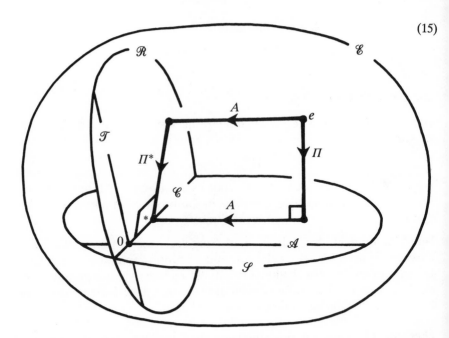

*Exercises*

1. The dual error distribution $Q$, defined by (1), has mean zero (so that, by (2), $V^+$ is its variance).

2. When is $(I^+)^+ = I$?

3. If $I = e_1^2 + \ldots + e_n^2$ then

$$\sum_{i=1}^{n} I^+(e_i, e)e_i \equiv e \qquad \text{for } e \in \mathcal{R}_I, \qquad (16)$$

$$\sum_{i=1}^{n} I^+(e_i, e)I^+(e_i, f) \equiv I^+(e, f) \qquad \text{for } e, f \in \mathcal{R}_I. \qquad (17)$$

4. $I$ is a singular inner product on $\mathcal{V}$, and $\mathcal{L}$ is a proper subspace of $\mathcal{E}$. Show that the set of $I^+$-orthogonal projections of $e$ on $\mathcal{L}$ equals $(e + \mathcal{L}^{I^+}) \cap \mathcal{L}$, a coset of $\mathcal{L} \cap \mathcal{N}_{I^+}$ (cf. the final paragraph of §7).

5. (Continuation.) Suppose $\mathcal{L} + \mathcal{R}_I = \mathcal{E}$ and that $I^+$ is chosen so that $\mathcal{N}_{I^+} \subset \mathcal{L}$. Let $\mathring{e}$ denote the projection of $e$ on $\mathcal{L}$ parallel to $\mathcal{S}^+$, the $I^+$-orthogonal-complement in $\mathcal{R}_I$ of $\mathcal{S} =_{\text{def}} \mathcal{L} \cap \mathcal{R}_I$. Show that the set of $I^+$-orthogonal projections of $e$ on $\mathcal{L}$ is equal to $e + \mathcal{N}_{I^+}$. Show also that $\mathcal{L}^\square$ is a subspace of $\mathcal{S}^\square$ that is an $I$-orthogonal complement of $(S^+)^\square$. Draw a picture that shows the dual of the projection $e \to \mathring{e}$ as $v \to \mathring{v}$, the $I$-orthogonal projection of $v$ on $(v + \mathcal{R}_{I^+}) \cap (\mathcal{S}^+)^\square$. Show that $\mathring{e}$ is more

simply characterizable by the identity

$$[\mathring{e}, v] = [e, v] - [e, \mathring{v}]$$

where $\mathring{v}$ is the $I$-orthogonal projection of $v$ on $\mathscr{L}^{\square}$.

6. (Continuation.) Investigate the possibility of a version of Theorem 8 for $I$ singular and $\mathscr{L} = \mathscr{R}_U$ that would deliver the projection $\mathring{e}$. Show that, when $\mathscr{L}$ is not a subspace of $\mathscr{R}_I$, we cannot simply replace $I^{-1}$ in (12) by $I^+$. What is $\Pi e$ when this replacement *is* made? (An equivalent, statistically inspired version of this problem will be explored in §30.)

7. (i) Why does the geological example (iii) of §2 imply a singular variance?

(ii) Observations have been made in a two-way layout involving two factors:

$$A \text{ at levels } A_1, \ldots, A_a,$$
$$B \text{ at levels } B_1, \ldots, B_b.$$

One observation was made at each of the $ab$ combinations $(A_i, B_j)$ $i = 1, \ldots, a$, $j = 1, \ldots, b$. Define the $p = (a + 1)(b + 1)$ variables, as follows: 'grand mean', 'main effect of $A_i$', $i = 1, \ldots, a$, 'main effect of $B_j$', $j = 1, \ldots, b$, 'interaction at $(A_i, B_j)$', $i = 1, \ldots, a$, $j = 1, \ldots, b$, with the customary interpretations in mind. Identify the constraints on realizable evaluators in the associated $p$-dimensional $\mathscr{E}$, and show that these lie in an $ab$ dimensional subspace of $\mathscr{E}$.

(iii) In an experiment to compare a new drug, $T$, with a currently-used one, $C$, in the treatment of a chronic condition, 100 similar patients are labelled from 1 to 100 completely at random. The 50 patients with labels 1 to 50 are then given $T$ for one year, the rest $C$. With $p = 100$, set

$$v_i = \text{'progress after 1 year for the patient labelled } i\text{'}$$

$i = 1, \ldots, 100$. The probability model for $[x, v_i]$ is

$$[x, v_i] = \theta_T + \delta_{r(i)}, \qquad i = 1, \ldots, 50$$
$$= \theta_C + \delta_{r(i)}, \qquad i = 51, \ldots, 100,$$

where $\theta_T$, $\theta_C$ are unknown scalar parameters, $r$ is a random permutation of $1, \ldots, 100$, and $\delta_1, \ldots, \delta_{100}$, with $\delta_1 \leqslant \delta_2 \leqslant \ldots \leqslant \delta_{100}$ and $\Sigma\, \delta_i = 0$, are unknown fixed scalar differences between the patients. Identify the error space $\mathscr{F}$ and find its dimension.

8. Show that a necessary and sufficient condition for eqn (14) to be reducible to the commutation $A\Pi = \Pi A$ is that $\mathscr{S}^{I^{-1}} \subset \mathscr{R}$.

9. Given a non-singular inner product $I$ on $\mathscr{V}$ and a subspace $\mathscr{S} \subset \mathscr{E}$, let $\Pi$ denote $I^{-1}$-orthogonal projection onto $\mathscr{S}$. Show that

(i) $I^\Pi$ agrees with $I$ on $I^{-1}\mathscr{S}$, both as inner product and as associated linear transformation,

(ii) for any choice of $\mathcal{R}_{(I^{\Pi})^+}$, $(I^{\Pi})^+$ agrees with $I^{-1}$ on $\mathcal{S}$ as inner product.

For what choice of $\mathcal{R}_{(I^{\Pi})^+}$ does the agreement in (ii) extend to the associated linear transformations?

## §12. Partial ordering of inner products

The variance of an affine transformation of a randomly distributed evaluator is a transformed inner product, as defined in §10, which suggests that the usual partial ordering of inner products may play a fundamental role in the optimizations needed for the solution of the statistical problems outlined in §4.

**Definition 1**  Given two inner products $I_1$ and $I_2$ on $\mathcal{V}$, we write $I_1 \leqslant I_2$ to mean that $I_1(v, v) \leqslant I_2(v, v)$ for all $v$.

**Definition 2**  Of two inner products $I_1$ and $I_2$ on $\mathcal{V}$, $I_1$ is the *smaller* if $I_1 \leqslant I_2$ and $I_1(v, v) < I_2(v, v)$ for some $v \neq 0$.

In line with Definition 2, $I_1$ is said to be a *minimum inner product* in a given set $\mathcal{I}$ of inner products that include $I_1$, if $I_1(v, v) = \min_{I \in \mathcal{I}} I(v, v)$ for every $v \in \mathcal{V}$, i.e. $I_1 \leqslant I$ for all $I$ in $\mathcal{I}$.

Analogous definitions apply to the comparison of inner products on $\mathcal{E}$. The following theorem gives a useful criterion for $I_1 \leqslant I_2$, in terms of the evaluation function of Definition 2 of §9.

**Theorem 1**  $I_1 \leqslant I_2$ if and only if $[I_1, J] \leqslant [I_2, J]$ for all non-singular inner products $J$ on $\mathcal{E}$.

**Proof**  From §9, for $J = \int v^2 \, d\beta$,

$$[I_2, J] - [I_1, J] = \int \{I_2(v, v) - I_1(v, v)\} \, d\beta,$$

which gives us the 'only if'. For the converse, with $I_1 = \int e^2 \, d\alpha_1$, and $I_2 = \int e^2 \, d\alpha_2$, we use the alternative integral representation to obtain

$$[I_2, J] - [I_1, J] = \int J(e, e)(d\alpha_2 - d\alpha_1). \tag{1}$$

Non-negativity of the right-hand side of (1) for all non-singular $J$ is a sufficient condition for $\int e^2 \, d\alpha_1 \leqslant \int e^2 \, d\alpha_2$ or $I_1 \leqslant I_2$. To see this, take $J = v_0^2 + \gamma J_0$ where $\gamma$ is a positive scalar and $J_0$ is a fixed non-singular inner product. Then

$$\int J(e, e)(d\alpha_2 - d\alpha_1) = I_2(v_0, v_0) - I_1(v_0, v_0) + \gamma([I_2, J_0] - [I_1, J_0])$$

and letting $\gamma \to 0$ implies $I_2(v_0, v_0) \geqslant I_1(v_0, v_0)$, where $v_0$ is arbitrary.  □

The evaluation function $[e, v]$ is the pivotal element in the proof of the following well-known result.

**Theorem 2**   If $I_1$ and $I_2$ are non-singular inner products on $\mathcal{V}$ and $I_1 \leqslant I_2$, then $I_2^{-1} \leqslant I_1^{-1}$.

**Proof**   (Suggested by Wang Jinglong.) For arbitrary $e$, let $v = I_2^{-1}e$. By Exercise 7 of §7,

$$[e, v]^2 \leqslant I_1^{-1}(e, e)I_1(v, v) \leqslant I_1^{-1}(e, e)I_2(v, v) = I_1^{-1}(e, e)I_2^{-1}(e, e).$$

But $[e, v] = I_2^{-1}(e, e)$, whence $I_2^{-1}(e, e) \leqslant I_1^{-1}(e, e)$.    □

A crucial property of the partial ordering, apart from its connection with variance minimization, is that it is inherited by shadow, marginal and transformed inner products. thus, in the terminology of §10,

$$I_1 \leqslant I_2 \Rightarrow \begin{cases} I_1^A \leqslant I_2^A & (A \text{ linear, } \mathcal{E} \rightarrow \mathcal{E}) \\ I_1^{\mathcal{T}} \leqslant I_2^{\mathcal{T}} & (\mathcal{T} \text{ a subspace of } \mathcal{E}) \\ I_1^B \leqslant I_2^B & (B \text{ linear, } \mathcal{E} \rightarrow \mathcal{G}). \end{cases} \qquad (2)$$

Perhaps the most important inequality of inner products for our purposes is one that is implicit in eqn (7) of §9 and eqn (4) of §11.

**Theorem 3**   (*Orthogonal Shadow Inequality*) For an inner product $I$ on $\mathcal{V}$ and $\Pi$ the $I^{-1}$- or $I^+$-orthogonal projection onto $\mathcal{S}$, a subspace of $\mathcal{R}_I$, we have

$$I^{\Pi} \leqslant I. \qquad (3)$$

The Orthogonal Shadow Inequality can be put to work immediately in the proof of another important theorem, concerning the minimization of transformed inner products when the transformation involved is constrained in the following way.

**Definition 2**   Given $B$ linear, $\mathcal{E} \rightarrow \mathcal{G}$, and $T$ linear, $\mathcal{S} \rightarrow \mathcal{G}$, where $\mathcal{S}$ is a subspace of $\mathcal{E}$, we say that $B$ *agrees with* $T$ on $\mathcal{S}$ if $Bs = Ts$ for all $s \in \mathcal{S}$.

**Theorem 4**   (*Transformed Inner Product Minimization.*) Let $I$ be a given inner product on $\mathcal{V}$ and suppose that $B$, linear $\mathcal{E} \rightarrow \mathcal{G}$, agrees on $\mathcal{S}$ with a fixed $T$, linear $\mathcal{S} \rightarrow \mathcal{G}$, where $\mathcal{S}$ is a fixed subspace of $\mathcal{R}_I$. Then $I^B$ is minimized when, acting on $\mathcal{R}_I$, $B$ equals $T\Pi$, where $\Pi$ is $I^{-1}$- or $I^+$-orthogonal projection onto $\mathcal{S}$.

**Proof**   By the Orthogonal Shadow Inequality (3), if $I = \int e^2 \, d\alpha$ then $I \geqslant \int (\Pi e)^2 \, d\alpha$. Hence, by (2),

$$I^B = \int (Be)^2 \, d\alpha \geqslant \int (B\Pi e)^2 \, d\alpha = \int (T\Pi e)^2 \, d\alpha$$

with equality if, acting on $\mathcal{R}_I$, $B$ equals $T\Pi$.    □

**Corollary** If $B$ is linear, $\mathscr{E} \to \mathscr{E}$, and $Bs = s$ for $s \in \mathscr{S} \subset \mathscr{R}_I$ then the shadow inner product $I^B$ is minimized when, acting on $\mathscr{R}_I$, $B$ equals $\Pi$.

**Proof** In Theorem 4, set $\mathscr{G} = \mathscr{E}$ and $T = 1$. $\qquad\qquad\qquad\qquad$ □

*Exercises*

1. Relaxing the condition in Theorem 2 that $I_1$ and $I_2$ be non-singular, show that $I_1 \le I_2$ implies $\mathscr{R}_{I_1} \subset \mathscr{R}_{I_2}$ and that, given $I_1 \le I_2$, a necessary and sufficient condition for $I_2^+ \le I_1^+$ is that $\mathscr{R}_{I_1} = \mathscr{R}_{I_2}$ and $I_1^+, I_2^+$ be chosen so that $\mathscr{N}_{I_1^+} = \mathscr{N}_{I_2^+}$.

2. Prove that, if $I$ is non-singular, the minimizing $B$ in the Corollary to Theorem 4 is unique. Is the same true in Theorem 4 itself? What can be said when $I$ is singular?

3. Construct an alternative 'duality proof' of Theorem 4 by showing firstly that, for fixed $g' \in \mathscr{G}'$,

$\quad$ (i) $\{B'g' : B \text{ agrees with } T \text{ on } \mathscr{S}\} = T'g' + \mathscr{S}^{\square}$,
$\quad$ (ii) $I^B(g', g')$ is then minimized when $B'g'$ is $I$-orthogonal to $\mathscr{S}^{\square}$.
(*Hint*: For the subcase of singular $I$, use Theorem 6 of §11.)

# 3
# Optimizations

In this chapter, the coordinate-free apparatus now at hand will be applied to some important problems of statistical optimization. In particular, the Orthogonal Shadow Inequality of §12 will be found to provide a smooth treatment of coordinate-free variants of the celebrated results of Gauss (1823) concerning best, linear, unbiased estimators. The Corollary of §12 will then be deployed for an equally facile resolution of prediction problems, both classical and Bayesian.

## §13. Gauss's linear model

The ground has been already prepared in §4 for a coordinate-free rendering of the most renowned of the classical problems of statistics— estimation and prediction for what may be called *Gauss's linear model*. When stated in coordinate-free terms, this model is, tersely,

$E(x) = \mu \in \mathscr{L}$, a known linear subspace of $\mathscr{E}$, and
$\mathrm{Var}(x) = V = \sigma^2 V_0$, where $V_0$ is known but the           (1)
scalar $\sigma^2$ may not be known.

Recall that $f =_{\mathrm{def}} x - \mu$ is the error vector and that the support of $f$ is the error subspace $\mathscr{F} = \mathscr{R}_V$. If $V$ is non-singular then $\mathscr{F} = \mathscr{E}$, but, if $V$ is singular, $\mathscr{F}$ is a proper subspace of $\mathscr{E}$. In considering any estimation based on $x$, it is essential to analyse, at an early stage, the implications of singularity of $V$. For, when $V$ is singular, the observation of $x$ reveals without error the particular coset $x + \mathscr{F} = \phi[x]$, say, of $\mathscr{E}/\mathscr{F}$ in which $x$ lies. Combining this revealed knowledge with the assertion of the model that $\mu \in \mathscr{L}$, we may conclude that, with probability 1, $\mu$ is in $\phi[x] \cap \mathscr{L}$. In other words, the values of $\mu$ that are probabilistically consistent with $x$ must lie in $\phi[x] \cap \mathscr{L}$. Symmetrically, the values of $x$, generated from the true value $\mu$, lie in $\phi[\mu]$ with probability 1. For statistical inference, only values of $x$ in $\mathscr{L} + \mathscr{F}$ need be considered: values outside $\mathscr{L} + \mathscr{F}$ have zero total probability for all values of $\mu \in \mathscr{L}$. Unless $\mathscr{L} \subset \mathscr{F}$, the partition $\mathscr{E}/\mathscr{F}$ divides the problem of estimation of $\mu$, or of a marginal parameter $\tau[\mu]$, into similar separate problems that we are free to analyse separately.

In contrast, when $V$ is non-singular, the observation of $x$ places no restriction on the assertion $\mu \in \mathscr{L}$, in the sense that $\mu$ can take any value in $\mathscr{L}$ and there are probability distributions with variance $V$ that give

43

positive probability to the observed $x$. In other words, the set of values of $\mu$ that can be probabilistically consistent with $x$ is the whole of $\mathscr{L}$, whatever $x$ is observed.

Clearly, a case has been made for logical distinctions, and perhaps procedural differences, in the treatment of the singular and non-singular cases. The latter is the simpler of the two and will be examined first.

*Exercise*

1. For the Gauss linear model, show that the value of a variable $v$ has expectation zero, whatever the value of $\mu$, if and only if $v \in \mathscr{L}^{\square}$.

## §14. Non-singular variance

We start with a simple theorem that plays a crucial role in our version of the classical Gauss theory, and whose inspiration is the statistical procedure for improving estimators known as Rao–Blackwellization (Rao 1945).

**Theorem 1**  Let $\mathscr{G}$ be any finite-dimensional vector space and $\tilde{\gamma} = g + Gx$ (with $g \in \mathscr{G}$ and $G$ linear $\mathscr{E} \to \mathscr{G}$), any affine transformation from $\mathscr{E}$ to $\mathscr{G}$. For Gauss's linear model ((1) of §13), let $\hat{\mu}$ denote the $V^{-1}$-orthogonal projection of $x$ on $\mathscr{L}$ and write $\hat{\gamma} = g + G\hat{\mu}$. Then, for all $\mu \in \mathscr{L}$,

(i)  $E(\tilde{\gamma}) = E(\hat{\gamma})$,
(ii) $\mathrm{var}(\tilde{\gamma}) \geqslant \mathrm{var}(\hat{\gamma})$.

**Proof**  $\hat{\mu} = \mu + \Pi f$ where $\Pi$ is $V^{-1}$-orthogonal projection onto $\mathscr{L}$. Hence $E(\hat{\mu}) = \mu = E(x)$, so that (i) obtains. For (ii), we have, by the Orthogonal Shadow Inequality (3) of §12, $\mathrm{var}(x) = V \geqslant V^\Pi = \mathrm{var}(\hat{\mu})$ whence $\mathrm{var}(Gx) \geqslant \mathrm{var}(G\hat{\mu})$.  $\square$

The theorem is interpretable as one on estimation, if $\tilde{\gamma}$ and $\hat{\gamma}$ are regarded as competing estimators of some fixed vector $\gamma$ in $\mathscr{G}$, no matter what $\gamma$ is. It tells us that, uniformly in $\mu \in \mathscr{L}$, the estimation bias, $E(\tilde{\gamma}) - \gamma$, is unchanged and that the *mean square error* ( = variance + (bias)$^2$) does not increase, if we replace $x$ by $\hat{\mu}$ in any affine estimator. In consequence, we can state

**Theorem 2**  *(Gauss Reduction)* If the criterion of accuracy of estimation is mean square error then we may reduce the class of affine estimators to

$$\{\hat{\gamma} = g + G\hat{\mu} : g \in \mathscr{G}, G \text{ linear } \mathscr{E} \to \mathscr{G}\} \qquad (1)$$

Any estimator in the reduced class (1) will be called a *Gauss estimator*. Borrowing freely from the language of decision theory, the class (1) could be described as 'complete within the class of affine estimators'. Note that Theorem 2 does not implicate unbiasedness in the justification of

$V^{-1}$-orthogonal projection onto $\mathscr{L}$ as the first step in estimation. Where the criterion of unbiasedness *will* enter is in the choice of $\hat{\gamma}$ from the class (1). We will see that involving unbiasedness at that stage in the argument may determine $\hat{\gamma}$ uniquely. Of course, the framework for the unbiasedness property remains to be defined.

The simplest example is undoubtedly the one in which $\mathscr{G} = \mathscr{L}$ and $\gamma \equiv \mu$. Then $\hat{\gamma}$ will be unbiased if and only if $g + G\mu \equiv \mu$, $\mu \in \mathscr{L}$, whence, by (1), $\hat{\gamma} \equiv \hat{\mu}$. We may thus conclude that $\hat{\mu}$ is the minimum variance, affine, unbiased estimator of $\mu$. An expression for the minimized variance may be obtained from Theorem 8 of §11. With $I = V$ and $U$ any inner product with range $\mathscr{L}$, we get

$$\begin{aligned} \mathrm{var}(\hat{\mu}) = \Pi V \Pi' &= U(UV^{-1}U)^{+}UV^{-1}VV^{-1}U(UV^{-1}U)^{+}U \\ &= U(UV^{-1}U)^{+}U, \end{aligned} \tag{2}$$

using (6) of §11.

Another example that does not take us outside $\mathscr{E}$ is given by $\mathscr{G} = \mathscr{E}/\mathscr{T}$, where $\mathscr{T}$ is some proper subspace of $\mathscr{E}$ with $\mathscr{T} \cap \mathscr{L} \neq \{0\}$ or $\mathscr{L}$, and $\gamma = \tau =_{\mathrm{def}} \tau[\mu] = \mu + \mathscr{T}$, a *'marginal parameter'*. For this, $\hat{\tau} = g + G\hat{\mu}$ will be unbiased if and only if $g + G\mu \equiv \tau[\mu]$, $\mu \in \mathscr{L}$, whence $\hat{\tau} \equiv \tau[\hat{\mu}]$. Hence $\tau[\hat{\mu}]$ is the minimum variance, affine, unbiased estimator of $\tau(\mu)$: the minimized variance is the marginal inner product

$$\mathrm{var}(\hat{\mu})^{\mathscr{T}} = (U(UV^{-1}U)^{+}U)^{\mathscr{T}}.$$

The picture for the estimation of $\tau$ is shown in Fig. 3. The general case of unrestricted $\mathscr{G}$ deserves a separate, formal statement:

$$\tag{3}$$

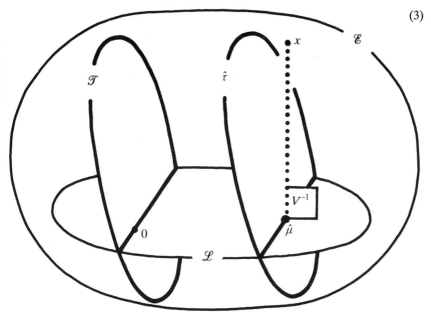

**Theorem 3**   (*Gauss Optimality*) If $\gamma \equiv G_1\mu$, $\mu \in \mathscr{L}$, where $G_1$ is given and linear, $\mathscr{L} \to \mathscr{G}$, the unique, unbiased Gauss estimator is $\hat{\gamma} = G_1\mu$, which is a minimum variance, affine, unbiased estimator of $\gamma$.

**Proof**   For unbiasedness, $g + G\mu \equiv G_1\mu$, whence $\hat{\gamma} \equiv G_1\hat{\mu}$. The 'minimum variance' is guaranteed by Theorem 2.   □

The transformation $\gamma \equiv G_1\mu$ may be regarded as an example of a *parametrization* of the Gauss model: we will give a full account of such parametrizations in Chap. V.

*Exercises*

1. Construct an alternative proof of the optimality of $\hat{\gamma}$ in Theorem 3, based on Theorem 4 of §12.
2. Establish the identity in $x$ and $v$

$$[\hat{\mu}, v] \equiv [x, \mathring{v}]$$

where $\mathring{v}$ is the $V$-orthogonal projection of $v$ on $V^{-1}\mathscr{L}$.

## §15. Singular variance

In view of the distinctions drawn in §13, the case of singular $V$ is, in general, the union of essentially disjoint subproblems. (We exclude the cases (*i*) $\mathscr{L} \subset \mathscr{F}$, which yields to the formal substitution of $\mathscr{E}$ for $\mathscr{F}$ and application of the methods of §14, and (*ii*) $\mathscr{F} \subset \mathscr{L}$, which has an intrinsic singularity leading to the trivial $\hat{\mu} \equiv x$.)

We therefore consider estimators $\tilde{\gamma}$ of the form

$$\tilde{\gamma} = g_{\phi[x]} + G_{\phi[x]}x \tag{1}$$

where, as in §13, $\phi[x] = x + \mathscr{F}$ and where, for each value $\phi \in \mathscr{E}/\mathscr{F}$ of the subscript, $g_\phi \in \mathscr{G}$ and $G_\phi$ is linear, $\mathscr{E} \to \mathscr{G}$.

**Definition**   Any estimator of the form (1) will be said to be *subaffine*.

The following extension of Theorem 1 of §14 is obtained by applying the singular case of the Orthogonal Shadow Inequality of §12:

**Theorem**   Let $\hat{\mu}$ denote the $V^+$-orthogonal projection of $x$ on $\phi[x] \cap \mathscr{L}$ and define $\hat{\gamma} = g_{\phi[x]} + G_{\phi[x]}\hat{\mu}$. Then, for all $\mu \in \mathscr{L}$,

(i)  $E(\tilde{\gamma}) = E(\hat{\gamma})$,
(ii)  $\operatorname{var}(\tilde{\gamma}) \geq \operatorname{var}(\hat{\gamma})$.

**Proof**   For each $\mu \in \mathscr{L}$, we have $x \in \phi[\mu]$ and hence $\phi[x] = \phi[\mu]$, with probability 1. Now $\hat{\mu} = \mu + \Pi f$ where $\Pi$ is $V^+$-orthogonal projection onto $\mathscr{S} = \mathscr{F} \cap \mathscr{L}$. So $E(\hat{\mu}) = \mu = E(x)$ whence (i) obtains. The Orthogonal

Shadow Inequality then applies within $\mathcal{F}$ to give

$$\text{var}(x) = \text{var}(f) = V \geqslant V^\Pi = \text{var}(\hat{\mu})$$

whence $\text{var}(\bar{\gamma}) \geqslant \text{var}(\hat{\mu})^{G_{\phi[\mu]}} = \text{var}(\hat{\gamma}).$  $\qquad\qquad\square$

The statement of the Gauss Reduction Theorem of §14 can now be modified, quite straightforwardly, to countenance restriction to the class of subaffine estimators

$$\{\hat{\gamma} = g_{\phi[x]} + G_{\phi[x]}\hat{\mu} : g_\phi \in \mathcal{G}, \ G_\phi \text{ linear } \mathcal{E} \to \mathcal{G}, \ \phi \in \mathcal{E}/\mathcal{F}\}. \qquad (2)$$

Most of the rest of §14 is likewise transferable, with minor changes. Thus, for the example in which $\mathcal{G} = \mathcal{L}$ and $\gamma \equiv \mu$, the estimator $\hat{\gamma}$ will be unbiased if and only if

$$g_{\phi[\mu]} + G_{\phi[\mu]}\mu \equiv \mu, \quad \mu \in \mathcal{L}. \qquad (3)$$

Now $\phi[\hat{\mu}] = \phi[x]$, so that (3) implies $\hat{\gamma} \equiv \hat{\mu}$, as was to be expected. An expression for $\text{var}(\hat{\mu})$, analogous to $U(UV^{-1}U)^+U$, will be obtainable once the promise made in Exercise 6 of §11 has been fulfilled.

The picture for Gauss Reduction in the singular case is shown in Fig. 4.

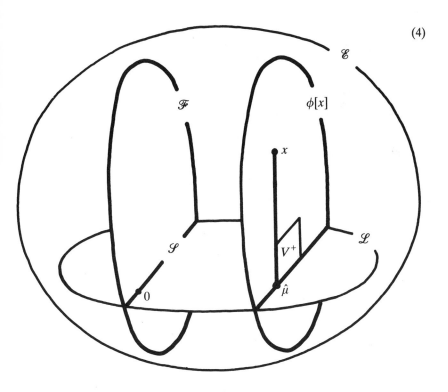

(4)

*Exercise*

1. For Exercise 7(iii) of §11, show that $V$ is known up to proportionality. Find a suitable $V_0$.

## §16. Error estimation

Suppose now that our concern is the estimation of $G_2 f =_{\mathrm{def}} G_2(x - \mu)$, where $G_2$ is given and linear, $\mathscr{E} \to \mathscr{G}$. One does not need to be much of a gambler to predict that $G_2(x - \hat{\mu})$ will be optimal in some sense. Moreover, it seems likely that this prediction could be verified by some modification of the methods of §14 and §15.

As it turns out, we can just as easily address the more general problem of estimation of the vector in $\mathscr{G}$ of the mixed type given by

$$\gamma(x) =_{\mathrm{def}} G_1\mu + G_2(x - \mu) \tag{1}$$

where $G_1$ is given and linear, $\mathscr{L} \to \mathscr{G}$.

For the case in which $V$ is non-singular, we start with the class of affine estimators of $\gamma(x)$

$$\{\tilde{\gamma} = g + Gx : g \in \mathscr{G}, \ G \text{ linear } \mathscr{E} \to \mathscr{G}\} \tag{2}$$

The analogue of Theorem 2 of §14 is that, if expectations and variances of the estimation error are our only concern, the class (2) may be reduced without loss to

$$\{\hat{\gamma} = g + G\hat{\mu} + G_2(x - \hat{\mu}) : g \in \mathscr{G}, \ G \text{ linear } \mathscr{L} \to \mathscr{G}\}. \tag{3}$$

This is because, for all $\mu \in \mathscr{L}$,

(i)  $E(\tilde{\gamma} - \gamma(x)) = E(\hat{\gamma} - \gamma(x))$,
(ii) $\mathrm{var}(\tilde{\gamma} - \gamma(x)) \geq \mathrm{var}(\hat{\gamma} - \gamma(x))$.

(Inequality (ii) is equivalent to

$$\mathrm{var}(G - G_2)x \geq \mathrm{var}(G - G_2)\hat{\mu},$$

which is an implication of the Orthogonal Shadow Inequality of §12.)

If we then ask for a $\hat{\gamma}$ from (3) that is unbiased for $\gamma(x)$ in the sense that $E(\hat{\gamma} - \gamma(x)) \equiv 0$, it is necessary that $g = 0$ and $G = G_1$, giving

$$\hat{\gamma} = G_1\hat{\mu} + G_2(x - \hat{\mu}) \tag{4}$$

as the affine estimator of $\gamma(x)$ whose error has minimum variance among those with zero expectation of error.

Estimator (4) is just (1) with $\mu$ estimated by $\hat{\mu}$. we have therefore overfulfilled our initial prediction for the special case $G_1 = 0$, namely that we do best to leave the $x$ in $G_2(x - \mu)$ unchanged and plug in the ubiquitous $\hat{\mu}$ in place of the unknown $\mu$.

For the case of singular variance, generalization of the results of §15 starts with the class of subaffine estimators of $\gamma(x)$

$$\{\hat{\gamma} = g_{\phi[x]} + G_{\phi[x]}x : g_\phi \in \mathcal{G}, \; G_\phi \text{ linear } \mathcal{E} \to \mathcal{G}, \; \phi \in \mathcal{E}/\mathcal{F}\}. \tag{5}$$

In analogy with (3), it can be shown that reduction may be made to the class

$$\{\hat{\gamma} = g_{\phi[x]} + G_{\phi[x]}\hat{\mu} + G_2(x - \hat{\mu}) : g_\phi \in \mathcal{G}, \; G_\phi \text{ linear } \mathcal{L} \to \mathcal{G}, \; \phi \in \mathcal{E}/\mathcal{F}\} \tag{6}$$

where $\hat{\mu}$ is the $V^+$-orthogonal projection of $x$ on $\phi[x] \cap \mathcal{L}$. The imposition of unbiasedness then gives (4), once again, as a subaffine estimator of $\gamma(x)$ whose error has minimum variance among those with zero expectation of error.

We have been able to use the results of $V^{-1}$-orthogonal and $V^+$-orthogonal projection of $x$ for estimation in the Gauss model because, although $V = \sigma^2 V_0$ is supposed known only up to proportionality, recognition of orthogonality does not require knowledge of $\sigma^2$. However, the value of $\sigma^2$ does come into the inferences based on the estimators thus obtained. The unbiased estimation of $\sigma^2$ is therefore our next concern.

**Definition 1** The *residual sum of squares* inner product for the Gauss linear model ((1) of §13) is defined by

$$RSS = (x - \hat{\mu})^2. \tag{7}$$

**Theorem 1**

$$E(RSS) = V^{1-\Pi}$$

where $\Pi$ is $V^{-1}$-orthogonal projection onto $\mathcal{L}$ when $V$ is non-singular, and $V^+$-orthogonal projection onto $\mathcal{F} \cap \mathcal{L}$, when $V$ is singular.

**Proof** The residual vector $x - \hat{\mu} = (1 - \Pi)x$ has zero expectation and variance $V^{1-\Pi}$. ∎

By Exercise 1 of §10 or its analogue for singular $I$, we have $V^{1-\Pi} = V - V^\Pi$. When $V$ is non-singular, the representation of $\text{var}(\hat{\mu})$ of §14 gives, additionally,

$$V^{1-\Pi} = V - U(UV^{-1}U)^+U.$$

**Theorem 2** An unbiased estimator of $\sigma^2$ is

$$\frac{V_0^{-1}(x - \hat{\mu}, x - \hat{\mu})}{\dim \mathcal{E} - \dim \mathcal{L}} \quad \text{when } V_0 \text{ is non-singular,} \tag{8}$$

$$\frac{V_0^+(x - \hat{\mu}, x - \hat{\mu})}{\dim \mathcal{F} - \dim(\mathcal{F} \cap \mathcal{L})} \quad \text{when } V_0 \text{ is singular.} \tag{9}$$

**Proof**

(i)
$$E\dot{V}_0^{-1}(x - \hat{\mu}, x - \hat{\mu}) = E[(x - \hat{\mu})^2, V_0^{-1}] = [V^{1-\Pi}, V_0^{-1}]$$
$$= [V^{1-\Pi}, V^{-1}]\sigma^2 = (\dim \mathscr{E} - \dim \mathscr{L})\sigma^2$$

by Theorem 1 of §10, since $1 - \Pi$ is $V^{-1}$-orthogonal projection onto $\mathscr{L}^{V^{-1}}$ which has dimensionality $\dim \mathscr{E} - \dim \mathscr{L}$.

(ii) Likewise, $EV_0^+(x - \hat{\mu}, x - \hat{\mu}) = [V^{1-\Pi}, V^+]\sigma^2 = (\dim \mathscr{F} - \dim(\mathscr{F} \cap \mathscr{L}))\sigma^2$ by Theorem 10 of §11. □

Note that $V_0^{-1}(x - \hat{\mu}, x - \hat{\mu})$ and $V_0^+(x - \hat{\mu}, x - \hat{\mu})$ can be written, somewhat more simply, as $[RSS, V_0^{-1}]$ and $[RSS, V_0^+]$ respectively.

## §17. Equivalent variances

For any given variance $V$ in the Gauss linear model, it is natural to consider what other values of $V$, if any, would leave the Gauss estimator $\hat{\mu}$ unchanged. Some knowledge about this class of *equivalent variances* might be especially useful in cases where $V$ is not fully specified. Putting the matter more symmetrically, we are therefore curious about possible equivalence classes of $V$, within which $\hat{\mu}$ has the same form.

We start by defining and characterizing the equivalence classes for non-singular variances, before considering the modifications needed for the singular variance case.

**Definition 1** Two non-singular variances $V_1$, $V_2$ are $\mathscr{L}$-*equivalent* if the $V_1^{-1}$- and $V_2^{-1}$-orthogonal projections onto $\mathscr{L}$ are identical.

**Theorem 1** Non-singular variances $V_1$, $V_2$ are $\mathscr{L}$-equivalent if and only if $V_1^{-1}\mathscr{L} = V_2^{-1}\mathscr{L}$ or, equivalently,

$$V_1\mathscr{L}^\square = V_2\mathscr{L}^\square. \tag{1}$$

**Proof** $\mathscr{L}$-equivalence $\Leftrightarrow \mathscr{L}^{V_1^{-1}} = \mathscr{L}^{V_2^{-1}} \Leftrightarrow V_1\mathscr{L}^\square = V_2\mathscr{L}^\square$ (by Theorem 2 of §6), while $\mathscr{L}^{V_1^{-1}} = \mathscr{L}^{V_2^{-1}} \Leftrightarrow V_1^{-1}\mathscr{L} = V_2^{-1}\mathscr{L}$ because $\mathscr{L}^{V_i^{-1}} = (V_i^{-1}\mathscr{L})^\square$, $i = 1, 2$). □

The dependence of $\hat{\mu}$ on $x$ and $V\mathscr{L}^\square$ alone is represented in Fig. 2. Condition (1) is also sufficient for equivalence when either or both of $V_1$ and $V_2$ are singular. However, in this case, we will adopt a more general definition of equivalence.

**Definition 2** Variances $V_1$, $V_2$ with error subspaces $\mathscr{F}_1$, $\mathscr{F}_2$, respectively, are $\mathscr{L}$-*equivalent on* $\mathscr{M}$, where $\mathscr{M}$ is a subspace with $\mathscr{L} \subset \mathscr{M} \subset \mathscr{E}$, $\mathscr{M} \subset \mathscr{L} + \mathscr{F}_1$ and $\mathscr{M} \subset \mathscr{L} + \mathscr{F}_2$, if, for every $x \in \mathscr{M}$, $\hat{\mu}_1 = \hat{\mu}_2$, where $\hat{\mu}_i$ is the estimate $\hat{\mu}$ of §15 with $V = V_i$, $i = 1, 2$.

(2)

(3)

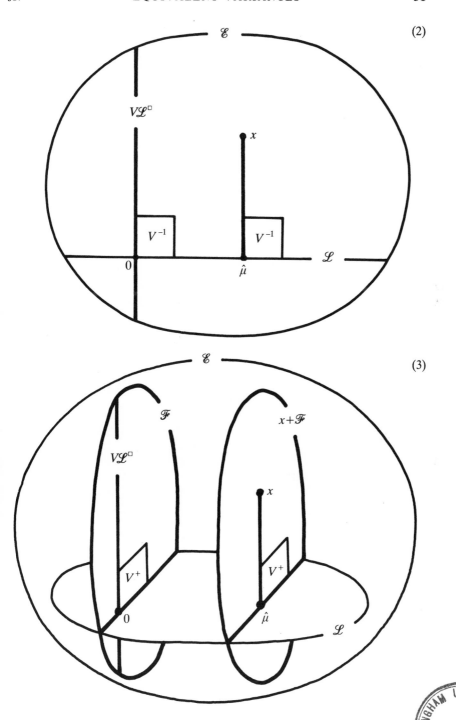

With this definition, our modification of Theorem 1 is

**Theorem 2**  Variances $V_1$, $V_2$ are $\mathscr{L}$-equivalent on $\mathscr{M}$ if and only if

$$\mathscr{M} \cap V_1 \mathscr{L}^{\square} = \mathscr{M} \cap V_2 \mathscr{L}^{\square}. \tag{4}$$

**Proof**  By Exercise 1 of §5, $V_i \mathscr{L}^{\square} = V_i(\mathscr{F}_i \cap \mathscr{L})^{\square}$, which, by Theorem 6 of §11 with $I = V_i$ and $\mathscr{S} = \mathscr{F}_i \cap \mathscr{L}$, equals the $V_i^+$-orthogonal complement of $\mathscr{F}_i \cap \mathscr{L}$ in $\mathscr{F}_i$. The estimate $\hat{\mu}_i$ in $\mathscr{L}$ is uniquely determined, for $x \in \mathscr{M}$, by the condition $x - \hat{\mu}_i \in V_i \mathscr{L}^{\square}$. So condition (4) is sufficient. If it were not also necessary, there would be a vector $x_0$ in $\mathscr{M}$ and $V_1 \mathscr{L}^{\square}$ but not in $V_2 \mathscr{L}^{\square}$ (or the same with $V_1$, $V_2$ interchanged.) For $x = x_0$, we would then have $\hat{\mu}_1 = 0 \neq \hat{\mu}_2$, a contradiction.  □

Figure 3 shows the dependence of $\hat{\mu}$ on $x$ and $V \mathscr{L}^{\square}$ in the singular case, and should serve to clarify the import of Definition 2 and Theorem 2. Because of the dependence on $\mathscr{M}$, the 'equivalence' of Theorem 2 does not generate equivalence classes. However, it seems better to leave matters as they stand since, for instance, condition (4) allows $V_1$ and $V_2$ to be recognized as equivalent on $\mathscr{L} + \mathscr{F}_1$, even if $\mathscr{F}_2$ is strictly larger than $\mathscr{F}_1$.

*Exercises*

1. If $V_2$ is non-singular, show that $V_1$ and $V_2$ are $\mathscr{L}$-equivalent for all $\mathscr{M}$ satisfying the conditions of Definition 2 if and only if $V_1 \mathscr{L}^{\square} \subset V_2 \mathscr{L}^{\square}$ or, equivalently, $V_1 V_2^{-1} \mathscr{L} \subset \mathscr{L}$.
2. (Continuation) Show that if, in addition, $V_1 V_2^{-1} \mu \equiv \mu$, $\mu \in \mathscr{L}$, then

$$\text{var}(\hat{\mu}_2 \mid V = V_2) = \text{var}(\hat{\mu}_2 \mid V = V_1)$$

(cf. Neuwirth 1982). (*Hint*: Use §14 (2) and §11 (12).)

## §18. Multivariate least-squares

The concept of equivalent variances is at the core of the multivariate generalization of the usual geometric treatment of univariate least-squares estimation.

The structure of the generalization is obtainable as a specialization of $\mathscr{E}$, $V$ and $\mathscr{L}$ in the Gauss linear model. Suppose $\mathscr{E} = \mathscr{Y}_1 \times \ldots \times \mathscr{Y}_n$, where $\mathscr{Y}_1, \ldots, \mathscr{Y}_n$ are replicas of the same finite-dimensional vector space $\mathscr{Y}$. We may write $\mathscr{E} = \mathscr{Y}^n$ for short. Suppose also that uncorrelated random vectors $y_1, \ldots, y_n$ are generated on $\mathscr{Y}$ with means $\eta_1, \ldots, \eta_n$ respectively and common non-singular variance $W$. For $i = 1, \ldots, n$, we then transfer $y_i$, $\eta_i$ and $W$ to $\mathscr{Y}_i$, and make the identifications $x = (y_1, \ldots, y_n)$ and $\mu = (\eta_1, \ldots, \eta_n)$.

The dual $\mathscr{V}$ of $\mathscr{E}$ may be equated with $\mathscr{U}_1 \times \ldots \times \mathscr{U}_n$, where $\mathscr{U}_i$ is a replica of $\mathcal{U}$, the dual of $\mathcal{Y}$. This identification is made by defining

$$[x, v] \equiv [y_1, u_1] + \ldots + [y_n, u_n]$$

where $v = (u_1, \ldots, u_n)$. The non-singular variance dualtor $V$ for $x$, that corresponds to the constant non-singular variance $W$ for each of the uncorrelated $y_1, \ldots, y_n$, is defined by

$$Vv \equiv (Wu_1, \ldots, Wu_n) \tag{1}$$

and we may write $V = (W, \ldots, W)$. The dual variance is $V^{-1} = (W^{-1}, \ldots, W^{-1})$, acting in the same component-wise way on $x$ as $V$ does on $v$.

The third ingredient in the specialization is the subspace of means, $\mathcal{L} \subset \mathcal{E}$. With an eye on Theorem 1 of §17 and its potential significance for the case in which the present $W$ is completely unknown, it would be nice if we could characterize the type of $\mathcal{L}$ for which $V^{-1}\mathcal{L}$ is the same for all non-singular $W$. The following condition on $\mathcal{L}$, that we will describe as the *scalar model*, is immediately seen to be *sufficient* for this invariance.

**Theorem 1**  If

$$\mathcal{L} = \left\{ \mu = (\eta_1, \ldots, \eta_n) : \eta_i = \sum_{j=1}^{r} x_{ij}\theta_j, \; x_{ij} \text{ a fixed scalar}, \; \theta_j \in \mathcal{Y} = \mathcal{Y}_i, \right.$$
$$\left. i = 1, \ldots, n, j = 1, \ldots, r \right\} \tag{2}$$

then $V^{-1}\mathcal{L}$ is the same for all non-singular $W$.

**Proof**

$$V^{-1}\mu = \left( \sum_{j=1}^{r} x_{ij}\phi_j \right)$$

where $\phi_j = W^{-1}\theta_j$, $j = 1, \ldots, r$. The range of values of $(\phi_1, \ldots, \phi_r)$ is the same for all $W$.                                                  □

We will consider the *necessity* of the scalar model (2), once the statistical case for being interested in the model has been established. Theorem 1, together with Theorem 1 of §17, informs us that the Gauss estimator $\hat{\mu}$, the uniquely defined $V^{-1}$-orthogonal projection of $x$ on $\mathcal{L}$, is the same for all non-singular $W$. The statistical relevance of this finding becomes clearer when $(\hat{\eta}_1, \ldots, \hat{\eta}_n) =_{\text{def}} \hat{\mu}$ is given a 'least-squares' characterization that shows explicitly the absence of any dependence on $W$. This characterization is a special case of

**Theorem 2**  $\bar{\mu} = (\bar{\eta}_1, \ldots, \bar{\eta}_n)$ minimizes $V^{-1}(x - \mu, x - \mu)$ for $\mu$ in any given subset of $\mathcal{E} = \mathcal{Y}^n$, for every non-singular $V = (W, \ldots, W)$ with action (1), if and only if it minimizes the 'deviation sum of squares'

$$D =_{\text{def}} \sum_{i=1}^{n} (y_i - \eta_i)^2. \tag{3}$$

**Proof** By Theorem 1 of §12, minimization of $D$ is equivalent to that of $[D, W^{-1}]$ for all non-singular $W$. But

$$[D, W^{-1}] = \sum_{i=1}^{n} W^{-1}(y_i - \eta_i, y_i - \eta_i) = V^{-1}(x - \mu, x - \mu). \qquad \square$$

**Corollary** $\hat{\mu} = (\hat{\eta}_1, \ldots, \hat{\eta}_n)$ is the 'least-squares' estimator in that it (uniquely) minimizes (3) for $\mu \in \mathcal{L}$.

The picture, Fig. 4, that best summarizes all this is almost indistinguishable from the one in $R^n$ that is customarily used to portray univariate least squares. The definition of $V^{-1}$ allows us to represent $\mathscr{E} = \mathscr{Y}^n$ as the direct sum of orthogonal subspaces $\mathscr{Y}_1, \ldots, \mathscr{Y}_n$. The fact that $\mathscr{Y}_1, \ldots, \mathscr{Y}_n$ may now be multidimensional does not compromise the simplicity that gives the picture such wide appeal in the univariate case. Thus, for $n = 2$, $r = 1$, we have

(4)

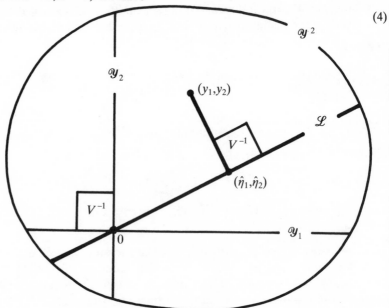

having established that we are actually analysing some version of 'multivariate least-squares', there is therefore a case for returning to the question of the necessity of the scalar model. The answer is in the affirmative.

**Theorem 3** The scalar model (2) is necessary for $V^{-1}\mathscr{L}$ to be the same for all non-singular $W$.

**Proof** The sets

$$\mathscr{S}_{1i} = \{\eta_i : (\eta_1, \ldots, \eta_n) \in \mathscr{L}\}, \qquad i = 1, \ldots, n,$$

are all subspaces of $\mathcal{Y}$. Since $V^{-1}\mathcal{L} = \{(W^{-1}\eta_1, \ldots, W^{-1}\eta_n): \mu \in \mathcal{L}\}$, the independence of $V^{-1}\mathcal{L}$ on $W$ means that $\mathcal{S}_{1i}$ is either $\{0\}$ or $\mathcal{Y}$ for all $i = 1, \ldots, n$. If all are $\{0\}$, then (2) obtains trivially. If not, suppose, without loss of generality, that $\mathcal{S}_{11} = \mathcal{Y}$.

Now consider the subspaces

$$\mathcal{S}_{2i} = \{\eta_i : (0, \eta_2, \ldots, \eta_n) \in \mathcal{L}\}, \qquad i = 2, \ldots, n,$$

each of which is likewise either $\{0\}$ or $\mathcal{Y}$. If all are $\{0\}$, then it follows that (2) obtains with $\theta_1 = \eta_1$, $x_{11} = 1$ and $x_{ij} = 0$ for $j \neq 1$. If not all $\mathcal{S}_{2i}$ are $\{0\}$, suppose without loss of generality that $\mathcal{S}_{22} = \mathcal{Y}$ and proceed to consider

$$\mathcal{S}_{3i} = \{\eta_i : (0, 0, \eta_3, \ldots, \eta_n) \in \mathcal{L}\}, \qquad i = 3, \ldots, n,$$

and so on, until, at the $(r + 1)$th stage either all $\mathcal{S}_{r+1,i}$ are empty or all are $\{0\}$.

At this point, $\mathcal{L}$ may be seen to be isomorphic with $\{(\eta_1, \ldots, \eta_r) \in \mathcal{Y}^r\}$. If $r = n$, (2) obtains trivially with $\theta_i = \eta_i$, $i = 1, \ldots, n$. If $r < n$ then the linearity of $\mathcal{L}$ implies that it is of the form

$$\left\{(\eta_1, \ldots, \eta_r, \sum_{j=1}^{r} A_{1j}\eta_j, \ldots, \sum_{j=1}^{r} A_{n-r,j}\eta_j) : (\eta_1, \ldots, \eta_r) \in \mathcal{Y}^r\right\} \quad (5)$$

where $\{A_{ij}\}$ are linear $\mathcal{Y} \to \mathcal{Y}$.

The independence of $V^{-1}\mathcal{L}$ on $W$ then implies that, for any nonsingular values $W_1$, $W_2$ of $W$ and all $(\eta_1, \ldots, \eta_r)$,

$$\left(W_1 W_2^{-1}\eta_1, \ldots, W_1 W_2^{-1}\eta_r, W_1 W_2^{-1} \sum_{j=1}^{r} A_{1j}\eta_j, \ldots, W_1 W_2^{-1} \sum_{j=1}^{r} A_{n-r,j}\eta_j\right) \in \mathcal{L}$$

and so is in (5). Hence

$$W_1 W_2^{-1} \sum_{j=1}^{r} A_{ij}\eta_j \equiv \sum_{j=1}^{r} A_{ij} W_1 W_2^{-1}\eta_j$$

identically in $(\eta_1, \ldots, \eta_r)$, whence $W_1 W_2^{-1}A_{ij} = A_{ij}W_1 W_2^{-1}$, or $W_1^{-1}A_{ij}W_1 = W_2^{-1}A_{ij}W_2$. Since $W_1$, $W_2$ are arbitrary non-singular inner products, it follows (see Exercise 4 of §7) that $A_{ij} = \alpha_{ij}1$ where $\alpha_{ij}$ is scalar. Substitution in (5) shows that, with restitution of the 'lost generality', $\mathcal{L}$ must be of scalar type. $\qquad\square$

*Exercises*

1. Show that, for any variance dualtor $V$ obeying (1) and for $\mathcal{L}$ given by (2) with $\mathbf{X} = (x_{ij})$ of full rank $r \leq n$, $V^{-1}$-orthogonal projection onto $\mathcal{L}$ acts on $(y_1, \ldots, y_n)$ as a 'column vector' by matrix multiplication with

the Kronecker-type matrix

$$M = \begin{bmatrix} \alpha_{11}1 & \alpha_{12}1 & \ldots & \alpha_{1n}1 \\ \vdots & \vdots & & \vdots \\ \alpha_{n1}1 & \alpha_{n2}1 & \ldots & \alpha_{nn}1 \end{bmatrix} \tag{6}$$

where $\alpha_{ij}$ is the scalar, $(i, j)$th element of the matrix $\mathbf{X}(\mathbf{X'X})^{-1}\mathbf{X'}$ and 1 is the identity on $\mathcal{Y}$. (*Hint*: Show that $M$ leaves $\mu \in \mathcal{L}$ unchanged and that

$$(1^d - M)\text{diag}(W^{-1}, \ldots, W^{-1})M = 0 \tag{7}$$

where $1^d$ is $\text{diag}(1, \ldots, 1)$.)

2. (Continuation) Show that, as linear transformation, $\mathcal{V} \to \mathcal{E}$, $\text{var}(\hat{\mu})$ acts on $(u_1, \ldots, u_n)$ as a 'column vector' by matrix multiplication with the Kronecker-type matrix

$$\text{var}(\hat{\mu}) = \begin{bmatrix} \alpha_{11}W & \ldots & \alpha_{1n}W \\ \vdots & & \\ \alpha_{n1}W & \ldots & \alpha_{nn}W \end{bmatrix}.$$

3. (Continuation) Defining $\hat{D} = \sum_{i=1}^{n} (y_i - \hat{\eta}_i)^2$, show that

$$E(\hat{D}) = (n - r)W.$$

(*Hint*: $(x - \hat{\mu})^2$ and $E(x - \hat{\mu})^2$, which is $\text{var}(x) - \text{var}(\hat{\mu})$, may be written in Kronecker form.)

4. Show that the $V$-orthogonal complement of $V^{-1}\mathcal{L}$ is also the same for all non-singular $W$. Relate this to the dual characterization of $\hat{\mu}$ in Exercise 2 of §14.

## §19. Alternative least-squares projections

There is, remarkably, another vector space in which multivariate least-squares estimation may be given a simple geometrical representation, with orthogonal projections of a quite different character. For the construction of this space, we turn to the scalars $\{x_{ij}\}$ in the definition of $\mathcal{L}$ in the scalar model (2) of §18:

$$E(y_i) = \eta_i = \sum_{j=1}^{r} x_{ij}\theta_j, \qquad i = 1, \ldots, n. \tag{1}$$

Write $\mathbf{x}_i$ for the $i$th *row* of the $n \times r$ matrix $\mathbf{X} = (x_{ij})$, and $z_i$ for the vector $(y_i, \mathbf{x}_i) \in \mathcal{Z} =_{\text{def}} \mathcal{Y} \times R^r$. For any fixed $\theta_1, \ldots, \theta_r$ in (1), the vectors

$$\zeta_i =_{\text{def}} (\eta_i, \mathbf{x}_i) = E(z_i), \qquad i = 1, \ldots, n,$$

span a subspace $\zeta$ of $\mathcal{Z}$, dependent on $\theta_1, \ldots, \theta_r$ and with $\zeta \cap \mathcal{Y} = \{0\}$

(treating $\mathcal{Y}$ as a subspace of $\mathcal{Z}$). If rank $\mathbf{X} = r$ then $\zeta$ is complementary to $\mathcal{Y}$.

Treating the vectors $z_1, \ldots, z_n$ in $\mathcal{Z}$ as a 'sample', two sample-based inner products on $\mathcal{Z}'$, the dual of $\mathcal{Z}$, may be defined:

$$Z_0 = \sum_{i=1}^{n} z_i^2 \qquad (Uncorrected\ Sum\ of\ Squares) \qquad (2)$$

$$Z = \sum_{i=1}^{n} (z_i - \bar{z})^2 \qquad (Total\ Sum\ of\ Squares). \qquad (3)$$

Note that $\mathcal{R}_{Z_0}$ and $\mathcal{R}_Z$ are the subspaces of $\mathcal{Z}$ spanned by $\{z_i\}$ and $\{z_i - \bar{z}\}$ respectively.

We find an immediate application for $Z_0$—to provide yet another characterization of the Gauss, least-squares estimator $\hat{\mu} = (\hat{\eta}_1, \ldots, \hat{\eta}_n)$ of §18. In this, a crucial role will be played by the translates $\xi_1, \ldots, \xi_n$ of $\mathcal{Y}$ that contain the sample vectors $z_1, \ldots, z_n$ respectively:

$$\xi_i = z_i + \mathcal{Y} = \{(y, \mathbf{x}_i) : y \in \mathcal{Y}\}, \qquad i = 1, \ldots, n.$$

**Theorem 1** For the scalar model (1), $\zeta_i =_{\text{def}} (\hat{\eta}_i, \mathbf{x}_i)$ is the $Z_0^{-1}$-orthogonal projection of the origin on $\xi_i$ or, if $Z_0$ is singular, its $Z_0^+$-orthogonal projection on $\xi_i \cap \mathcal{R}_{Z_0}$.

**Proof** Let $\Pi$ be $Z_0^{-1}$-orthogonal projection onto $\mathcal{Y}$, or $Z_0^+$-orthogonal projection onto $\mathcal{Y} \cap \mathcal{R}_{Z_0}$. Write $\bar{\eta}_i = y_i - \Pi z_i$. We have to show that $\bar{\eta}_i \equiv \hat{\eta}_i$. If $B$, linear $\mathcal{Z} \to \mathcal{Z}$, is an identity on $\mathcal{Y}$, then by the Corollary of §12, we have

$$\sum_{i=1}^{n} (y_i - \bar{\eta}_i)^2 = Z_0^{\Pi} \leq Z_0^B. \qquad (4)$$

We are also free to define $B$ so that it annihilates $\zeta$ i.e. all the vectors $\zeta_i = (\eta_i, \mathbf{x}_i)$, $i = 1, \ldots, n$. Then, additionally,

$$Z_0^B = \sum_{i=1}^{n} (y_i - \eta_i)^2, \qquad (5)$$

and $\sum_{i=1}^{n} (y_i - \bar{\eta}_i)^2 \leq \sum_{i=1}^{n} (y_i - \eta_i)^2$. So, by the Corollary of §18, we have $\bar{\eta}_i \equiv \hat{\eta}_i$, since $(\bar{\eta}_1, \ldots, \bar{\eta}_n) \in \mathcal{L}$.     □

Figure 6 records the optimal projection in the non-singular case. Figure 7, for the singular case, in which $z_1, \ldots, z_n$ do not span $\mathcal{Z}$ is slightly more complex. An important special case of the scalar model (1) is when $\mathbf{X}$ has a column, $\mathbf{1}$, of equal (non-zero) elements. Without loss of generality, these equal values may be taken to be unity, so that we have what may be described as the location case. For simplicity, let us analyse it as the extension of (1) given by

$$E(y_i) = \eta_i = \theta_0 + \sum_{j=1}^{r} x_{ij}\theta_j, \qquad i = 1, \ldots, n. \qquad (8)$$

(6)

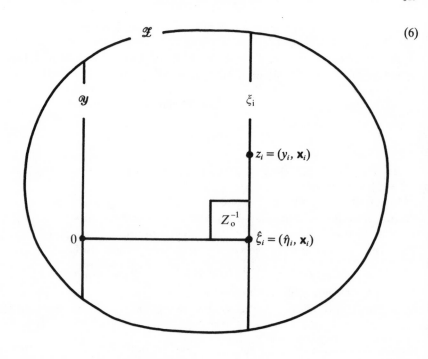

(7)

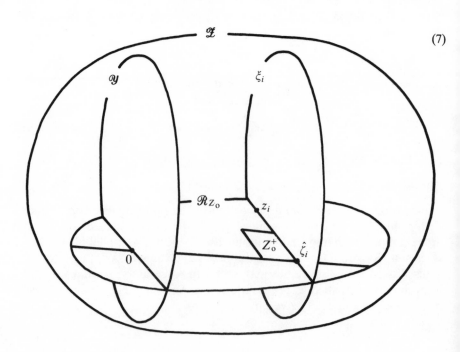

As a special case of the scalar model, the *location model* (8) could, of course, be treated as (1) was treated, using the appropriate $Z_0$. It is, however, statistically more insightful to bring the Total Sum of Squares inner product $Z$ into play. In order to do this, we will keep $\mathbf{x}_i = (x_{i1}, \ldots, x_{ir})$ and $z_i = (y_i, \mathbf{x}_i)$ as before, but adjust the definition of $\zeta_i$ to be $(\eta_i - \theta_0, \mathbf{x}_i)$, so that $\zeta_i$ is formally unchanged as

$$\left( \sum_{j=1}^{r} x_{ij}\theta_j, \mathbf{x}_i \right).$$

Then $\zeta_1, \ldots, \zeta_n$ still span a subspace $\zeta$. The translate

$$\zeta^t =_{\text{def}} \theta_0 + \zeta$$

includes the $n + 1$ vectors

$$\zeta_i^t =_{\text{def}} (\eta_i, \mathbf{x}_i) = E(z_i), \qquad i = 1, \ldots, n,$$

and

$$\theta_0 = \zeta^t \cap \mathcal{Y},$$

and is, in fact, their affine span.

**Theorem 2**   For the location model (8),

$$\hat{\zeta}_i^t =_{\text{def}} (\hat{\eta}_i, \mathbf{x}_i)$$

is the $Z^{-1}$-orthogonal projection of $\bar{z}$ on $\xi_i = z_i + \mathcal{Y}$ or, if $Z$ is singular, its $Z^+$-orthogonal projection on $\xi_i \cap (\bar{z} + \mathcal{R}_Z)$.

**Proof**   This is a straightforward extension of the proof of Theorem 1. Now, we write $\bar{\eta}_i = y_i - \Pi(z_i - \bar{z})$. As before, taking $B$ to be an identity on $\mathcal{Y}$ and to annihilate $\zeta$, we find, with a little help from Fig. 9,

$$\sum_{i=1}^{n} (y_i - \bar{\eta}_i)^2 = Z^\Pi \leqslant Z^B = \sum_{i=1}^{n} (B(z_i - \bar{z}))^2$$

$$\leqslant \sum_{i=1}^{n} (B(z_i - \bar{z}))^2 + n(B\bar{z} - \theta_0)^2 = \sum_{i=1}^{n} (Bz_i - \theta_0)^2$$

$$= \sum_{i=1}^{n} (y_i - \eta_i)^2.$$

The proof is completed by the argument used for Theorem 1.    □

The picture of the optimal projection for the case of $Z$ non-singular is given by a straightforward adaptation of Fig. 6, and need not be reproduced. As if to restore the pictorial balance, however, the singular case has two subcases that call for separate portraits. The distinction between them, to be exploited, hinges on whether or not $\mathcal{Y}$, as origin in

(9)

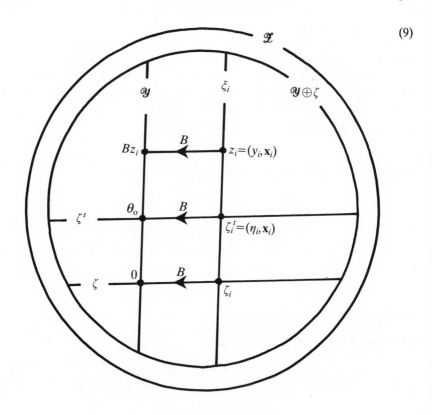

the vector space $\mathscr{X}/\mathscr{Y}$, satisfies the condition

$$\mathscr{Y} = \alpha_1\xi_1 + \ldots + \alpha_n\xi_n \text{ for some scalar } \alpha_1, \ldots, \alpha_n \text{ with } \sum \alpha_i = 1$$
(10)

i.e. whether or not $\mathscr{Y}$ is in the affine span of $\xi_1, \ldots, \xi_n$. When $Z$ is non-singular, (10) always obtains; for singular $Z$, a necessary and sufficient condition for (10) is

$$\text{rank}(\mathbf{1} \quad \mathbf{X}) = (\text{rank } \mathbf{X}) + 1.$$
(11)

When (10) is satisfied, application of eqn (16) of §11 with

$$\mathscr{X}/\mathscr{Y}, \ \{(\xi_i - \bar{\xi})/(n-1)^{\frac{1}{2}}\}, \qquad Z_{\xi\xi} =_{\text{def}} \sum (\xi_i - \bar{\xi})^2/(n-1),$$

$$-\bar{\xi} = \sum \alpha_i(\xi_i - \bar{\xi})$$

replacing $\mathscr{E}$, $\{e_i\}$, $I$ and $e$, respectively, establishes that we may take $\alpha_i$ in

(10) to be

$$\bar{\alpha}_i = \frac{1}{h} - \frac{1}{(n-1)} Z_{\xi\xi}^+(\xi_i - \xi, \xi), \qquad i = 1, \ldots, n, \tag{12}$$

(with $Z_{\xi\xi}^+$ standing for $Z_{\xi\xi}^{-1}$ when $Z_{\xi\xi}$ is non-singular).

**Definition 1** The *least-squares manifold* $\overset{\scriptscriptstyle t}{\zeta}{}^{t}$ is the affine span of $\overset{\scriptscriptstyle t}{\zeta}{}^{t}_1, \ldots, \overset{\scriptscriptstyle t}{\zeta}{}^{t}_n$.

**Definition 2**

$$\hat{\theta}_0 =_{\text{def}} \zeta^t \cap \mathcal{Y}. \tag{13}$$

**Theorem 3** $\hat{\theta}_0$ is non-null if and only if condition (10) obtains.

**Proof** This follows from the equivalence of the eqns (10) and

$$\zeta^t \cap \mathcal{Y} = \alpha_1 \zeta^t \cap \xi_1 + \ldots + \alpha_n \zeta^t \cap \xi_n, \tag{14}$$

an equivalence stemming from the linearity of the 1–1 correspondence between $\zeta^t \cap \xi$ and $\xi$, for $\xi$ such that $\zeta^t \cap \xi \neq \varnothing$.     □

**Theorem 4** If $Z$ is non-singular, then $(\hat{\theta}_0, \mathbf{0})$ is the $Z^{-1}$-orthogonal projection of $\bar{z}$ on $\mathcal{Y}$ or, if $Z$ is singular and condition (10) obtains, its $Z^+$-orthogonal projection on $\mathcal{Y} \cap (\bar{z} + \mathcal{R}_Z)$.

**Proof** Projection commutes with the affine combination of cosets of $\mathcal{Y}$ involved.     □

With all this set-theory, we are entitled to call for reassuring pictures, at least for the singular case. Firstly, the subcase in which (10) holds is shown in Fig. 15. The less interesting case of null $\hat{\theta}_0$ is the subject matter of Fig. 16. Vectors in $\bar{z} + \mathcal{R}_Z$ are all expressible as the sum of a vector in the affine space $\zeta^t$ and one in the vector subspace $\mathcal{Y}$. Hence

$$\hat{\theta}_0 = \zeta^t \cap \mathcal{Y} = \phi \Rightarrow (\bar{z} + \mathcal{R}_Z) \cap \mathcal{Y} = \varnothing.$$

This nullity, on top of that of $\hat{\theta}_0$, is indicated by the separated representation of $\mathcal{Y}$ in Fig. 16, living, as it must do, in higher dimensions than those of the minimal affine space that will accommodate the other structures.

   The following result about $\hat{\theta}_0$ as a potential estimator of $\theta_0$ is of interest because, as its Corollary shows, the arbitrariness of the origin of the vector $\mathbf{x}$ in the location model (8) allows us to take $\theta_0$ as representative of any interesting affine combination of $\eta_1, \ldots, \eta_n$. Concentration on $\hat{\theta}_0$ will also prepare the ground for the analysis of a superficially distinct problem in §22.

(15)

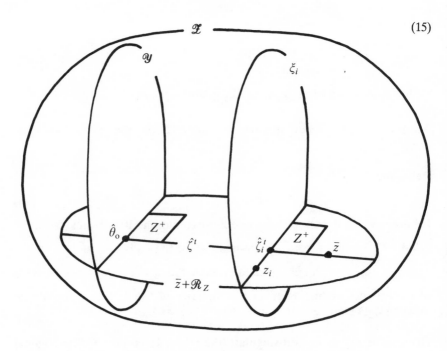

(16)

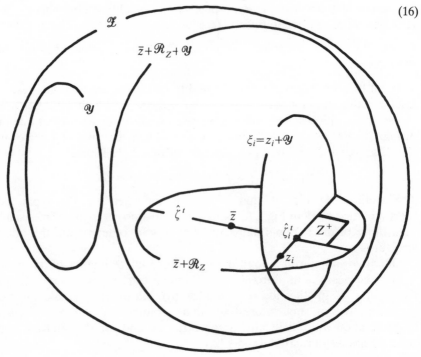

**Theorem 5**   Under condition (10),

$$\theta_0 = \alpha_1 \eta_1 + \ldots + \alpha_n \eta_n \tag{17}$$

$$\hat{\theta}_0 = \alpha_1 \hat{\eta}_1 + \ldots + \alpha_n \hat{\eta}_n \tag{18}$$

and $\hat{\theta}_0$ is the minimum variance, affine, unbiased estimator of $\theta_0$. Furthermore

$$\hat{\theta}_0 = \bar{y} - Z_{y\xi} Z_{\xi\xi}^+ \bar{\xi} = \tilde{\alpha}_1 y_1 + \ldots + \tilde{\alpha}_n y_n \tag{19}$$

where $Z_{y\xi} = \sum\limits_{i=1}^{n} (y_i - \bar{y})(\xi_i - \bar{\xi})/(n-1)$ and $\{\tilde{\alpha}_i\}$ are given by (12).

Moreover,

$$\text{var}(\hat{\theta}_0) = \left\{ \frac{1}{n} + \frac{1}{n-1} Z_{\xi\xi}^+(\bar{\xi}, \bar{\xi}) \right\} W. \tag{20}$$

**Proof**   Condition (10) implies

$$\zeta' \cap Y = \alpha_1 \zeta' \cap \xi_1 + \ldots + \alpha^n \zeta' \cap \xi_n$$

which gives (17). Similar use of $\xi'$ gives (18). The stated optimality of $\hat{\theta}_0$ then derives from Theorem 3 of §14 with $\gamma = \theta_0$.

To establish the useful expressions for $\hat{\theta}_0$ in (19), we simply combine Theorem 4 with Theorem 7 of §11, replacing $\mathscr{E}$, $I$, $\mathscr{S}$, $\mathscr{T}$ in the latter by $\mathscr{X}$, $Z$, $\mathscr{Y}$, $\mathscr{X}/\mathscr{Y}$, respectively, and treating $\mathscr{X}$ as $\mathscr{Y} \oplus (\mathscr{X}/\mathscr{Y})$. The measure $\alpha$ in $I$ specializes to a measure of $1/(n-1)$ on each of the points $(y_i - \bar{y}) + (\xi_i - \bar{\xi})$, $i = 1, \ldots, n$. The required $Z^{-1}$ or $Z^+$-orthogonal projection of $\bar{z} = \bar{y} + \bar{\xi}$ is then given by eqn (9) of §11 as

$$\bar{y} - \left( \sum \frac{(y_i - \bar{y})(\xi_i - \bar{\xi})}{n-1} \right) \left( \sum \frac{(\xi_i - \bar{\xi})^2}{n-1} \right)^+ \bar{\xi} \tag{21}$$

which is the first form of $\hat{\theta}_0$ in (19). The second form then follows by use of the definition of $\{\tilde{\alpha}_i\}$ in eqn (12).

Finally, $\text{var}(\hat{\theta}_0) = \left( \sum\limits_{i=1}^{n} \tilde{\alpha}_i^2 \right) W$ and

$$\sum \tilde{\alpha}_i^2 = \frac{1}{n} + \frac{1}{(n-1)^2} \sum Z_{\xi\xi}^+ (\xi_i - \bar{\xi}, \bar{\xi})^2$$

$$= \frac{1}{n} + \frac{1}{n-1} \sum (Z_{\xi\xi}^+ ((\xi_i - \bar{\xi})/(n-1)^{\frac{1}{2}}, \bar{\xi}))^2$$

$$= \frac{1}{n} + \frac{1}{n-1} Z_{\xi\xi}^+ (\bar{\xi}, \bar{\xi})$$

by an application of eqn (17) of §11.   □

If we now simply change the origin in the vector-space $\mathscr{X}/\mathscr{Y}$ from $\mathscr{Y}$ to $\xi$, Theorem 5 acquires a more general formulation. Despite its coordinate-free disguise, the result should then have a more familiar appearance.

**Corollary**  If $\xi = (0, \mathbf{x}) + \mathscr{Y}$ is in the affine span of $\xi_1, \ldots, \xi_n$, then $\hat{\theta}$ defined by

$$(\hat{\theta}, \mathbf{x}) = \xi^t \cap \xi,$$

is the minimum variance, affine, unbiased estimator of

$$\theta =_{\text{def}} \theta_0 + \sum_{j=1}^{r} x_j \theta_j.$$

Moreover,

$$\hat{\theta} = \bar{y} + Z_{y\xi} Z_{\xi\xi}^{+}(\xi - \bar{\xi})$$

and

$$\text{var}(\hat{\theta}) = \left\{ \frac{1}{n} + \frac{1}{n-1} Z_{\xi\xi}^{+}(\xi - \bar{\xi}, \xi - \bar{\xi}) \right\} W.$$

*Exercises*

1. Check the validity of the dualtor notation in the following expression for $\hat{\eta}_i$, and its derivation as an application of Theorem 7 of §11 to the current Theorem 1:

$$\hat{\eta}_i = \left\{ \sum y_j(0, \mathbf{x}_j) \right\} \left\{ \sum (0, \mathbf{x}_j)^2 \right\}^{+}(0, \mathbf{x}_i). \tag{22}$$

Make the connection with the alternative representation obtainable from Exercise 1 of §18.

2. Defining $m$ to be dim $\mathscr{Y}$ and $q$ to be the rank of the matrix $\mathbf{X}$ in the general scalar model (1), state the conditions on the quadruple $(m, n, q, r)$ that determine

   (i)  whether or not $Z_0$ is necessarily singular,
   (ii) whether or not $\hat{D}$ (Exercise 3 of §18) is necessarily zero,
   (iii) whether or not $\hat{D}$ is necessarily singular, as an inner product on the dual of $\mathscr{Y}$.

3. Establish the equivalence of (10) and (11).

4. Prove the assertion about (12).

5. Does the optimality of $\hat{\theta}$ in the Corollary require that $W$ be non-singular?

## §20. Conditional estimation in Gauss's linear model

One component of the observation of $x$ in Gauss's linear model, as stated in (1) of §13, is the value $\lambda$ of the coset $\lambda[x] = x + \mathcal{L}$ in which $x$ lies. When the distribution of the error $f = x - \mu$ is completely known, $\lambda[x]$ is an ancillary statistic, having a distribution that is known and the same for all values of $\mu \in \mathcal{L}$. If nothing is known about the error distribution apart from $V_0$ then $\lambda[x]$ is no longer ancillary. There is, nonetheless, some case for considering the conditional properties of the class of subaffine estimators of the form

$$\bar{\mu} = a_{\lambda[x]} + B_{\lambda[x]} x \tag{1}$$

in which, for each $\lambda \in \mathcal{E}/\mathcal{L}$, $a_\lambda$ is in $\mathcal{E}$, and $B_\lambda$ is linear, $\mathcal{E} \to \mathcal{E}$. By 'conditional', we mean with respect to the conditional distribution of $x$ given $\lambda[x] = \lambda$, for general $\lambda$.

For the case of non-singular $V$, we will now show that, if the error distribution has a certain linearity property with respect to $\mathcal{L}$ and if $\bar{\mu}$ is required to be 'conditionally unbiased', then $\bar{\mu}$ is identical with the Gauss estimator $\hat{\mu}$. This invests the Gauss estimator with a conditional optimality not involving any variance minimization. A trivial result, perhaps, but one that serves to prepare the ground for the more interesting case of 'replication' in the next section.

**Definition 1** The error distribution is *linear in* $\mathcal{E}/\mathcal{L}$ if the conditional expectations

$$\{E(f \mid \lambda[f] = \lambda) : \lambda \in \mathcal{E}/\mathcal{L}\}$$

span a subspace complementary to $\mathcal{L}$. (To be rigorous, we should specify a definition of the conditional expectation, for $\lambda$ in the support of $\lambda[f]$. To cut a corner, suppose this has been done to the satisfaction of a measure theorist.)

**Theorem 1** If the error distribution is linear in $\mathcal{E}/\mathcal{L}$ then

$$E(f \mid \lambda[f] = \lambda) = O_\lambda,$$

where $O_\lambda$ is the $V^{-1}$-orthogonal projection of the origin on $\lambda$, i.e. the conditional expectations span $\mathcal{L}^{V^{-1}}$.

**Proof** Let $\Pi_c$ denote projection onto $\mathcal{L}$ parallel to the subspace of conditional expectations, and $\Pi$ the $V^{-1}$-orthogonal projection onto $\mathcal{L}$. Then

$$V^{\Pi_c} = E\{(\Pi_c f)^2\} = E_\lambda E_{f|\lambda}(f - E(f \mid \lambda))^2$$

(2)

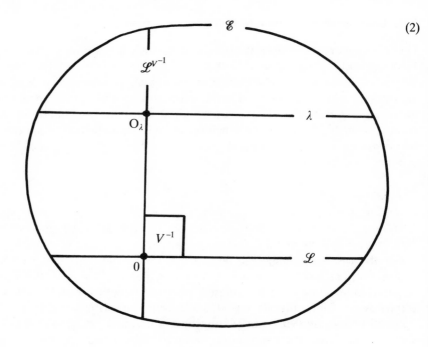

and

$$V^{\Pi} = E_\lambda E_{f \mid \lambda}(f - O_\lambda)^2.$$

By the Corollary of §12, $V^{\Pi} \leqslant V^{\Pi_c}$, while $E(f \mid \lambda)$ is known to be the value of $e \in \lambda$ minimizing $E_{f \mid \lambda}(f - e)^2$. Hence $E(f \mid \lambda) = O_\lambda$. $\qquad\square$

The structures for Theorem 1 are recorded in Fig. 2.

**Definition 2** $\bar\mu$, given by (1), is *conditionally unbiased given* $\lambda[x] = \lambda$ if

$$E(\bar\mu \mid \lambda[x] = \lambda) = \mu \qquad \text{for all } \mu \in \mathcal{L}.$$

**Theorem 2** If the error distribution is linear in $\mathcal{E}/\mathcal{L}$ then, given $\lambda[x] = \lambda$, the Gauss estimator $\hat\mu$ is the only conditionally unbiased, subaffine estimator of $\mu$.

**Proof** Let $\mu_\lambda$ denote the conditional expectation of $x$ given $\lambda[x] = \lambda$. Then, by Theorem 1, $\mu_\lambda = \mu + O_\lambda$ for $\mu \in \mathcal{L}$, while conditional unbiasedness of $\bar\mu$ means that

$$a_\lambda + B_\lambda \mu_\lambda \equiv \mu$$

for $\lambda \in \mathcal{E}/\mathcal{L}$ and $\mu \in \mathcal{L}$. Hence

$$a_{\lambda[x]} + B_{\lambda[x]}x \equiv x - O_{\lambda[x]} = \hat\mu. \qquad\square$$

### §21. Replication in Gauss's model

Consider a random sample $x_1, \ldots, x_n$ from the probability distribution $P$ in the Gauss linear model (1) of §13. For simplicity, we will consider only the case of non-singular $V$, known up to proportionality.

A generalization of the argument of §14, from $n = 1$ to general $n$, is obtainable by a straightforward '$n$-folding' of all the ingredients. Thus we create $\tilde{x} = (x_1, \ldots, x_n)$, put it in the product space $\tilde{\mathscr{E}} =_{\text{def}} \mathscr{E} \times \ldots \times \mathscr{E}$ and define

$$\tilde{\mathscr{L}} = \{\tilde{\mu} = (\mu, \ldots, \mu) : \mu \in \mathscr{L}\}$$

which is a subspace of $\tilde{\mathscr{E}}$. The dual of $\tilde{\mathscr{E}}$ is $\tilde{\mathscr{V}} = \mathscr{V} \times \ldots \times \mathscr{V}$ and $\text{var}(\tilde{x})$ is

$$\tilde{V} = (V, \ldots, V), \tag{1}$$

with dual

$$\tilde{V}^{-1} = (V^{-1}, \ldots, V^{-1}), \tag{2}$$

where it is understood that (1) and (2) act componentwise on $\tilde{\mathscr{V}}$ and $\tilde{\mathscr{E}}$ respectively. Application of the Gauss Reduction Theorem of §14 to this formulation tells us that optimality in the affine estimation of some $\gamma$ in a vector space $\mathscr{G}$ is to be found in the reduced class of affine estimators of the form $\hat{\gamma} = g + G\hat{\mu}$, where $\hat{\mu}$ is the $\tilde{V}^{-1}$-orthogonal projection of $\tilde{x}$ on $\tilde{\mathscr{L}}$. However,

$$\tilde{V}^{-1}(\tilde{x} - \tilde{\mu}, \tilde{x} - \tilde{\mu}) = \sum_{i=1}^{n} V^{-1}(x_i - \bar{x}, x_i - \bar{x}) + nV^{-1}(\bar{x} - \mu, \bar{x} - \mu),$$

whence $\hat{\tilde{\mu}} = (\hat{\mu}, \ldots, \hat{\mu})$ where $\hat{\mu}$ is the $V^{-1}$-orthogonal projection of $\bar{x}$ on $\mathscr{L}$.

That the Gauss Reduction Theorem should deliver the sample mean $\bar{x}$ as the operationally 'sufficient' statistic is not a statistically surprising outcome. The interest here lies in the mode of delivery.

The sample mean $\bar{x}$ is likewise involved in the generalization for a random sample of the conditional analysis of §20. The single condition $\lambda[x] = \lambda$ is now multiplied to the $n$-fold condition $\lambda[x_i] = \lambda_i$, $i = 1, \ldots, n$. Let $O_i$ denote the $V^{-1}$-orthogonal projection of the origin on $\lambda_i$, and $\mu_i$ the conditional expectation of $x_i$ given $\lambda[x_i] = \lambda_i$. Then, if the error distribution is linear in $\mathscr{E}/\mathscr{L}$, Theorem 1 of §20 dictates that

$$\mu_i = \mu + O_i, \qquad i = 1, \ldots, n.$$

Writing $y_i = x_i - O_i$, we see that $y_1, \ldots, y_n$ are conditionally independently distributed vectors in the same space $\mathscr{L}$ with common mean $\mu$.

**Definition 1** The error distribution is *homoscedastic in* $\mathscr{E}/\mathscr{L}$ if the

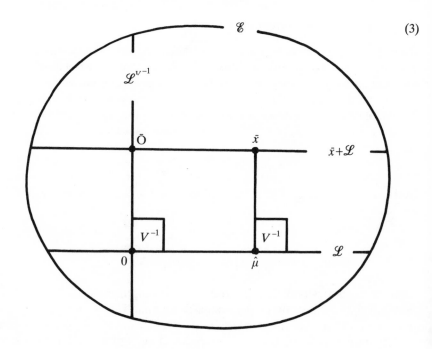

(3)

conditional variance $\mathrm{var}(f \mid \lambda[f] = \lambda)$ is the same for all $\lambda$. Denote this common conditional variance $V_c$.

With homoscedasticity, we can employ a strict analogue of the $n$-folding argument already considered, with $\mathscr{L}$ replacing $\mathscr{E}$, $(y_1, \ldots, y_n)$ replacing $(x_1, \ldots, x_n)$ and $V_c$ replacing $V$. It is immaterial that $\mathscr{L}$, that started as a proper subspace of the space, $\mathscr{E}$, containing the replications $x_1, \ldots, x_n$, now fills out the space in which the 'replications' $y_1, \ldots, y_n$ are distributed.

Another application of Gauss Reduction then requires that the starting point for affine estimation of $\mu$ be the $V_c^{-1}$-orthogonal projection on $\mathscr{L}$ of $\bar{y} = \bar{x} - \bar{O}$, where $\bar{O} = (O_1 + \ldots + O_n)/n$. Since $\bar{y}$ is already in $\mathscr{L}$, the projection is $\bar{x} - \bar{O}$ itself, which, as Fig. 3 illustrates, is nothing other than $\hat{\mu}$, the $V^{-1}$-orthogonal projection of $\bar{x}$ on $\mathscr{L}$. Thus, with linearity and homoscedasticity of the error structure, the conditional analysis has the same endpoint as the unconditional analysis. Applying Gauss Optimality (Theorem 3 of §14), we may conclude that, under the stated conditions, $G_1\hat{\mu}$ is the minimum variance, affine, conditionally unbiased estimator of a given parameter $\gamma = G_1\mu$, $\mu \in \mathscr{L}$. We note, for future comparison, that, under the same conditions, the conditional and unconditional variances of $\hat{\mu}$ are equal, with common value $V_c/n$.

## §22. Replication and estimated variance

What if, in the problem just considered, the variance $V$ were not known, i.e. not known up to proportionality? Can the present approach find anything to say in favour of the 'plug-in' estimator of $\mu$, given by the $S^{-1}$-orthogonal projection of $\bar{x}$ on $\mathscr{L}$ or, with rather less initial plausibility, by the $S^{+}$-orthogonal projection of $\bar{x}$ on $\mathscr{L} \cap (\bar{x} + \mathscr{R}_S)$ when the sample variance $S = \sum (x_i - \bar{x})^2/(n-1)$ is singular? The basis of the idea that it may is just that $S$ is an unbiased estimator of $V$.

Suppose that the error distribution, i.e. that of each $x_i - \mu$, is linear in $\mathscr{E}/\mathscr{L}$ (Definition 1 of §20) and that $\dim(\mathscr{E}/\mathscr{L}) = r$. Select a fixed basis $\{q_1, \ldots, q_r\}$ of $\mathscr{E}/\mathscr{L}$, and let $\phi_j$ be the conditional expectation of the error within coset $q_j$, $j = 1, \ldots, r$. Let $x_{i1}, \ldots, x_{ir}$ be the (scalar) coordinates of $\lambda_i = x_i + \mathscr{L}$, i.e.

$$\lambda_i = \sum_{j=1}^{r} x_{ij}q_j, \qquad i = 1, \ldots, n. \tag{1}$$

Then, in the notation of §21, the $V^{-1}$-orthogonal projection of the origin on $\lambda_i$ is

$$O_i = \sum_{j=1}^{r} x_{ij}q_j \cap \mathscr{L}^{V^{-1}} \tag{2}$$

in light of Theorem 1 of §20. Adding $\mu$ to both sides of (2) and setting $\phi_j = q_j \cap \mathscr{L}^{V^{-1}}$ gives

$$\mu_i = \mu + \sum_{j=1}^{r} x_{ij}\phi_j \tag{3}$$

for the conditional expectation of $x_i$ given $\lambda[x_i] = \lambda_i$. This temporary lapse into coordinates has revealed that we have the makings of an equivalent of the 'location' case of 'multivariate least-squares' in §19. To complete the connection, we make the additional assumption that the error distribution is homoscedastic in $\mathscr{E}/\mathscr{L}$ (Definition 1 of §21).

To make the equivalence absolutely clear, we take the risk of notational overload and define

(i)   $\phi_{j0} = q_j \cap \mathscr{M}_0$, $j = 1, \ldots, r$, where $\mathscr{M}_0$ is any fixed subspace of $\mathscr{E}$, complementary to $\mathscr{L}$,

(ii)   $\mu_{i0} = \sum_{j=1}^{r} x_{ij}\phi_{j0} = \lambda_i \cap \mathscr{M}_0$, $i = 1, \ldots, n$, $\qquad\qquad (4)$

(iii)   $y_i = x_i - \mu_{i0}$ and $\eta_i = \mu_i - \mu_{i0}$, $i = 1, \ldots, n$,

(iv)   $\theta_0 = \mu$, $\quad \theta_j = \phi_j - \phi_{j0}$, $j = 1, \ldots, r$.

Then $y_1, \ldots, y_n$ are independent vectors in $\mathscr{L}$, with conditional expectations given by the following close analogue of (8) of §19:

$$E(y_i \mid \lambda[x_i] = \lambda_i) = \eta_i = \mu_i - \mu_{i0} = \theta_0 + \sum_{j=1}^{r} x_{ij}\theta_j \tag{5}$$

with conditional variances all equal to $V_c$.

Finally, identifying the current $\mathscr{L}$ with the space $\mathscr{Y}$ of §19, we are almost ready to exploit the results of that section. However, we must ensure firstly that, given the linearity on $\mathscr{E}/\mathscr{L}$ of the error distribution, the current parameter of interest $\mu$ (equivalent to the $\theta_0$ and *not* the $\mu$ of §19) is determined, necessarily linearly, by $\eta_1, \ldots, \eta_n$. The work on $\theta_0$ in §19 informs us that a necessary and sufficient condition for this determination is that $\mathscr{L}$ (as origin in $\mathscr{E}/\mathscr{L}$) should lie in the affine span of $\lambda_1, \ldots, \lambda_n$, i.e.

$$\mathscr{L} = \alpha_1\lambda_1 + \ldots + \alpha_n\lambda_n \tag{6}$$

where $\alpha_1, \ldots, \alpha_n$ are scalars with $\alpha_1 + \ldots + \alpha_n = 1$. The parameter of interest is then

$$\mu = \sum \alpha_i\mu_i = \sum \alpha_i\eta_i. \tag{7}$$

Equation (7) can be derived algebraically from (3), (4), (5), and (6). However, it is more easily seen as part of the equivalence with the location case of §19, for which the unavoidable *discrepancies* of notation are:

| §19 | | Here |
|-----|---|------|
| $\mathscr{X}$ | $\sim$ | $\mathscr{E}$ |
| $\mathscr{Y}$ | $\sim$ | $\mathscr{L}$ |
| $z$ | $\sim$ | $x$ |
| $\theta_0$ | $\sim$ | $\mu$ |
| $\mu$ | $\sim$ | $(\eta_1, \ldots, \eta_n)$ |
| $\xi$ | $\sim$ | $\lambda$ |
| $W$ | $\sim$ | $V_c$ |

Orthogonal projections based on $S$, here, are equivalent to those based on $Z$ in §19. We can get our 'plug-in' estimator of $\mu$, $\hat{\mu}$ say, as

$$\hat{\mu} = \alpha_1\hat{\eta}_1 + \ldots + \alpha_n\hat{\eta}_n \tag{8}$$

whre $\hat{\eta}_i$ is the $\mathscr{L}$-component of the $S^{-1}$-orthogonal projection of $\bar{x}$ on $\lambda_i = x_i + \mathscr{L}$ or, if $S$ is singular, the $S^+$-orthogonal projection of $\bar{x}$ on

$\lambda_i \cap (\bar{x} + \mathcal{R}_S)$. Alternatively, $\hat{\mu}$ in (8) is the direct outcome of the same projections on $\mathcal{L}$ or $\mathcal{L} \cap (\bar{x} + \mathcal{R}_S)$, respectively.

The value of $\hat{\mu}$ is not affected by any non-uniqueness of the $\{\alpha_i\}$ in (6). Following (12) of §19, we may take

$$\alpha_i = \frac{1}{n} - \frac{1}{n-1} L^+ (\lambda_i - \bar{\lambda}, \bar{\lambda}), \qquad i = 1, \ldots, n,$$

where

$$L =_{\text{def}} \sum_{i=1}^{n} (\lambda_i - \bar{\lambda})^2 / (n-1).$$

So much for method! For the reassurance of optimality of $\hat{\mu}$, given that (6) obtains, the equivalence with §19 is equally serviceable. Without further analysis, we can conclude that $\hat{\mu}$ is the minimum (conditional) variance, affine in $(y_1, \ldots, y_n)$, (conditionally) unbiased estimator of $\mu$.

It may be written explicitly in the form

$$\hat{\mu} = \bar{y} - \frac{1}{n-1} \left\{ \sum_{i=1}^{n} (y_i - \bar{y})(\lambda_i - \bar{\lambda}) \right\} L^+ \bar{\lambda} \tag{9}$$

with conditional variance given by

$$\text{var}(\hat{\mu}) = \left( \frac{1}{n} + \frac{1}{n-1} L^+(\bar{\lambda}, \bar{\lambda}) \right) V_c. \tag{10}$$

Figure 11 illustrates the relationship between $\hat{\mu}$, based on the estimate $S$

$$\tag{11}$$

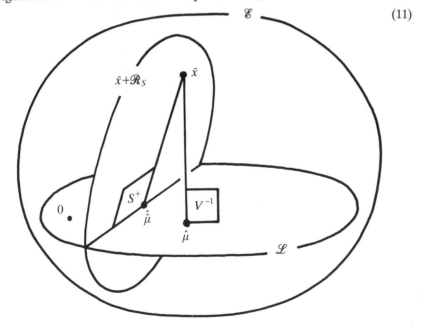

of $V$, and the estimator $\hat{\mu}$ of §21, based on knowledge of $V$ up to proportionality. Expression (10) for $\text{var}(\hat{\mu})$ quantifies the price that has to be paid for the estimation of $V$, in comparison with the generally smaller value $V_c/n$ of $\text{var}(\hat{\mu})$ (see the end of §21), and reveals how that price depends on $\bar{\lambda}$. Note that for $\bar{\lambda} = \mathscr{L}$, which is the origin of $\mathscr{E}/\mathscr{L}$, $L^+(\bar{\lambda}, \bar{\lambda}) = 0$ and the two variances are then equal.

*Exercises*

1. By means of examples with small $p$ and $n$ values, show that the probability that $\mathscr{L}$ lies in the affine span of $\lambda_1, \ldots, \lambda_n$ can take any value in $[0, 1]$.

2. Investigate the conditions under which $\hat{\mu}$ is the unique, affine in $(y_1, \ldots, y_n)$, conditionally unbiased estimator of $\mu$, making variance minimization redundant.

## §23. Prediction

Partial observation of a randomly distributed evaluator $x$ raises the problem of prediction of the unobserved (hidden or missing) part, from the part that has been observed. The term 'prediction' is justified on the grounds that the observed part may be the past record of $x$, the remainder of $x$ representing the unobserved features. However, there is nothing to prevent us predicting a future revelation about the past!

Our coordinate-free model for partial observation is that $x$ is known to lie in some particular coset of a quotient space $\mathscr{E}/\mathscr{H}$. Letting $\mathscr{K}$ be some subspace complementary to $\mathscr{H}$, this knowledge about $x$ is equivalent to knowledge of $k$ in the component representation $x = h + k$. Prediction is then required for the 'hidden' component $h \in \mathscr{H}$. Two cases will be considered.

**Case 1**  If the probability distribution $P$ of $x$ were completely known, prediction would optimally be based on the conditional distribution of $x$, given $x \in k + \mathscr{H}$. However, if all that is known are the values of $\mu = E(x) = E(h) + E(k)$ and $V = \text{var}(x)$, the latter up to proportionality only, we may be content to use an affine predictor of $h$ of the form $\hat{h} = a + Bk$, where $a \in \mathscr{E}$ and $B$ is linear, $\mathscr{E} \to \mathscr{E}$. (It is convenient to think of $B$ as having domain $\mathscr{E}$ and being an annihilator of $\mathscr{H}$, rather than to restrict its action to its effective domain $\mathscr{K} \sim \mathscr{E}/\mathscr{H}$.) As the criterion of optimality, let us use the unconditional mean square error inner product. The variance $V$ will be taken to be non-singular: if it were not, we could make the accommodation for singularity, as in Theorem 7 of §11, without affecting the essence of the results that may now be stated.

**Theorem 1**    The optimal affine predictor of $h$ is

$$\hat{h} = E(h) + \text{cov}(h, k)\text{var}(k)^+(k - E(k)) \tag{1}$$

which has mean square error

$$E(h - \hat{h})^2 = \text{var}(h) - \text{cov}(h, k)\text{var}(k)^+\text{cov}(k, h).$$

**Proof**

$$E(h - a - Bk)^2 = E(h - E(h) - B\{k - E(k)\})^2 + \{a - E(h) + BE(k)\}^2$$
$$= V^A + \{a - E(h) + BE(k)\}^2,$$

where $Ax =_{\text{def}} h - Bk$. The only condition on $A$ is that $Ah \equiv h$, i.e. $A$ is an identity on $\mathcal{H}$. Hence, by the Corollary of §12, mean square error is minimized when $A$ is $V^{-1}$-orthogonal projection onto $\mathcal{H}$ and, at the same time, $a = E(h) - BE(k)$. Theorem 7 of §11 then gives

$$Ax = h - \text{cov}(h, k)\text{var}(k)^+ k,$$

whence $B = \text{cov}(h, k)\text{var}(k)^+$ and $\hat{h}$ is given by (1).

The minimized mean square error is

$$E\{h - E(h) - \text{cov}(h, k)\text{var}(k)^+(k - E(k))\}^2$$
$$= \text{var}(h) - \text{cov}(h, k)\text{var}(k)^+\text{cov}(k, h),$$

using some dualtor algebra including (6) of §11.        □

Figure 3 shows the prediction $\hat{x} = \hat{h} + k$ as the $V^{-1}$-orthogonal projection of $E(x)$ on the coset $k + \mathcal{H}$ ($=x + \mathcal{H}$) that is known to contain $x$. We note for future reference that $\hat{x}$ may be written in the alternative form

$$\hat{x} = E(x) + \text{cov}(x, k)\text{var}(k)^+(k - E(k)). \tag{2}$$

in which we see that $\hat{x}$ is an unbiased predictor of $x$ in the sense that their overall expectations are equal.

**Case 2**    Suppose now that knowledge of $\mu$ and $V$, non-singular, is that of Gauss's linear model—$\mu \in \mathcal{L}$ and $V$ known up to proportionality. There are two sub-cases to be distinguished, according to whether $\mathcal{H} \cap \mathcal{L} = \{0\}$ or not. (The further subcase $\mathcal{L} = \{0\}$ has been already covered by Case 1.)

We start with the sub-case $\mathcal{H} \cap \mathcal{L} = \{0\}$. To simplify the analysis without loss of generality, take $\mathcal{H}$ to be the $V^{-1}$-orthogonal complement of $\mathcal{H}$ in $\mathcal{E}$.

Figure 4 helps to clarify the problem, and to expose the lines of a simple solution. It does not represent faithfully all possible cases, in that it portrays $\mathcal{H} \oplus \mathcal{L}$ as a proper subspace of $\mathcal{E}$, and $\mathcal{H}$ and $\mathcal{L}$ as satisfying $\mathcal{H} \cap \mathcal{L} = \{0\}$. However, the reader will be able to check that all cases are

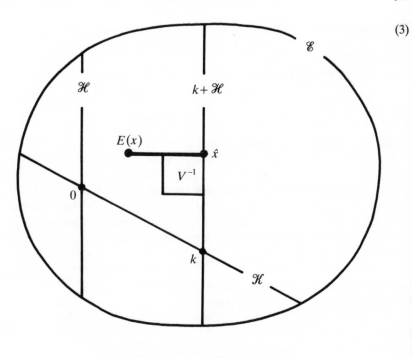

(3)

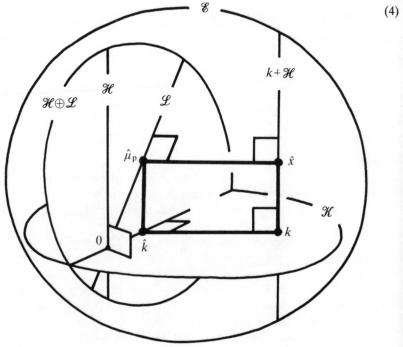

(4)

properly covered. In Fig. 4, $\mathscr{S} = (\mathscr{H} \oplus \mathscr{L}) \cap \mathscr{K}$, $\hat{k}$ is the $V^{-1}$-orthogonal projection of $k$ on $\mathscr{S}$, and $\hat{\mu}_{\mathrm{p}}$ is the vector in $\mathscr{L}$ that has $\mathscr{S}$-component $\hat{k}$ in its resolution $\hat{h} + \hat{k}$ into $\mathscr{H}$ and $\mathscr{S}$ components, which defines $\hat{h}$ too. Then $\hat{x}$, the $V^{-1}$-orthogonal projection of $\hat{\mu}_{\mathrm{p}}$ on the coset $k + \mathscr{H}$ in which $x$ is known to lie, has the resolution

$$\hat{x} = \hat{h} + \hat{k} + (k - \hat{k}),$$

where the three components are $V^{-1}$-orthogonal. Note that $\hat{x}$, thus defined, is a determined linear transformation of the known vector $k$.

It seems likely that $\hat{x}$ is an optimal predictor of $x$ in some sense. We show now that it is indeed optimal, in the sense that the prediction error

$$e_{\mathrm{p}} = x - \hat{x} = h - \hat{h}$$

has minimum variance for affine predictors whose prediction errors have zero expectation for all $\mu \in \mathscr{L}$. In fact, $\hat{x}$ has a broader optimality, involving the requirement of zero-expectation only for specified linear transformations of $e_{\mathrm{p}}$, analogous to $G_1\mu + G_2(x - \mu)$ in §16. However, this generalization will not be treated here.

One way of establishing the optimality of $\hat{x}$ is to define an alternative general affine predictor $\bar{x}$, which may be written in the form

$$\bar{x} = (a + Bk) + k, \tag{5}$$

and to choose $a$ and $B$ to minimize the prediction mean square error inner product

$$E(h - a - Bk)^2. \tag{6}$$

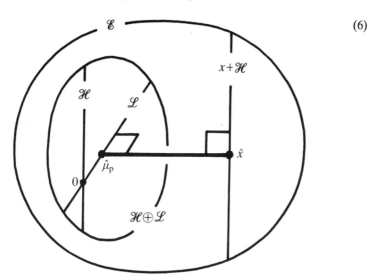

The reader is invited to pursue this method of proof as an Exercise: it yields the expected solution after

(i) an application of a prediction variant of the Gauss Reduction Theorem, implying that $a + Bk$ may always be replaced by $a + B\hat{k}$, and

(ii) the imposition of the condition of zero-expectation on the error of prediction.

However, inspection of Fig. 4 shows that $\mathcal{H}$ plays no essential role in the determination of $\hat{x}$. We could equally well have started with Fig. 6, which dispenses with $\mathcal{H}$, defining $\hat{\mu}_p$ and $\hat{x}$ as the vectors in $\mathcal{L}$ and $x + \mathcal{H}$, respectively, that minimize the $V^{-1}$-distance between these subsets of $\mathcal{E}$. (The subscript in $\hat{\mu}_p$ reminds us that we are dealing with the estimation of $\mu$ as a spin–off from the prediction of $x$.) Intuitively, a proof based on (6) *ought* to be simpler than one based on use of $\mathcal{H}$ in (4). That this is indeed the case rests on the observation that if, from the start, we had written our general affine predictor (5) in the form $\bar{x} = a + Bx$ then the fact that $B$ has to annihilate $\mathcal{H}$ means that, equivalently,

$$\bar{x} = a + B\hat{x}. \tag{7}$$

The expectation of the prediction error is then

$$E(x - \bar{x}) = -a - (B - 1)\mu \qquad \text{for } \mu \in \mathcal{L}.$$

Imposing the condition that the expectation be identically zero implies that $a = 0$ and that $B$ is an identity on $\mathcal{L}$. Then, by (7),

$$E(x - \bar{x})^2 = E\{x - \hat{x} + (1 - B)(\hat{x} - \hat{\mu}_p)\}^2 \geqslant E(x - \hat{x})^2 \tag{8}$$

establishing the optimality of $\hat{x}$. The *inequality* in (8) holds because $E(x - \hat{x})(\hat{x} - \hat{\mu}_p) = E(\hat{x} - \hat{\mu}_p)(x - \hat{x}) = 0$, derivable from an application of Theorem 2 of §9 with $\mathcal{H} \oplus \mathcal{L}$, $(\mathcal{H} \oplus \mathcal{L})^{V^{-1}}$, $V$ in place of $\mathcal{S}$, $\mathcal{T}$, $I$ (in which derivation $x - \hat{x}$ is the $\mathcal{H}$-component in $\mathcal{H} \oplus \mathcal{L}$ of the $V^{-1}$-orthogonal projection of $e = x - \mu$ on $\mathcal{H} \oplus \mathcal{L}$). Explicit formulae for $\hat{\mu}_p$ and $\hat{x}$ may be stated in terms of $V$ and inner product dualtors, $H$ and $U$, whose ranges are $\mathcal{H}$ and $\mathcal{L}$, respectively. Their derivation will be postponed until we have dealt with the necessary generalizations of the inverse transformations $H^+$ and $U^+$ of §11.

Now consider the sub-case $\mathcal{H} \cap \mathcal{L} \neq \{0\}$. The crucial difference that this makes is a structural one. When $\mathcal{H} \cap \mathcal{L} = \{0\}$, knowledge of the $\mathcal{H}$-component of $\mu \in \mathcal{L}$ serves to fix $\mu$. But, when $\mathcal{H} \cap \mathcal{L} \neq \{0\}$, there is non-uniqueness up to a coset of $\mathcal{H} \cap \mathcal{L}$. It is to be expected that this structural distinction will be somehow reflected in the solution of the associated prediction problem.

Dispensing with $\mathcal{H}$ and taking the data as the knowledge of $x + \mathcal{H}$, we start with the necessary modification of Fig. 6. This is shown in Fig. 9.

(9)

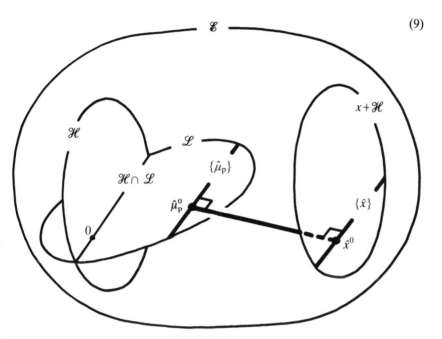

The pair of points, $\hat{\mu}_p \in \mathscr{L}$ and $\hat{x} \in x + \mathscr{H}$, that are $V^{-1}$-closest are now not uniquely defined. For, if we find one such pair, $\hat{\mu}_p^0$, $\hat{x}^0$ say, then the pair $\hat{\mu}_p^0 + e$, $\hat{x}^0 + e$ are equally close for any $e \in \mathscr{H} \cap \mathscr{L}$. In fact, $\{\hat{\mu}_p\} = \hat{\mu}_p^0 + (\mathscr{H} \cap \mathscr{L})$ and $\{\hat{x}\} = \hat{x}^0 + (\mathscr{H} \cap \mathscr{L})$, as illustrated.

The essential statistical action can be concisely represented in a *marginalization* of $\mathscr{E}$. Define

$$\delta[e] = e + (\mathscr{H} \cap \mathscr{L}),$$

the natural correspondence between $\mathscr{E}$ and $\mathscr{E}/(\mathscr{H} \cap \mathscr{L})$. Let $\eta = \delta[\mu]$ and $y = \delta[x]$. For any set $\mathscr{D} \subset \mathscr{E}$, define

$$\delta[\mathscr{D}] = \{\delta[e] : e \in \mathscr{D}\}.$$

The marginalization $\delta$ produces Fig. 10 in which

$$\hat{\eta}_p =_{\text{def}} \delta[\hat{\mu}_p], \qquad \hat{y} =_{\text{def}} \delta[\hat{x}],$$

independent of the choices of $\hat{\mu}_p$ and $\hat{x}$. The following lemma shows that $\hat{\eta}_p$ and $\hat{y}$ must play a key role in the marginalized problem.

**Lemma 1**   $\hat{\eta}_p$, $\hat{y}$ is the unique, $(V^\delta)^{-1}$-closest pair of vectors in $\delta[\mathscr{L}]$, $\delta[x + \mathscr{H}]$, respectively.

**Proof**   For notational convenience but without loss of generality, we may identify $\delta[\mathscr{E}]$ with some fixed subspace of $\mathscr{E}$ complementary to

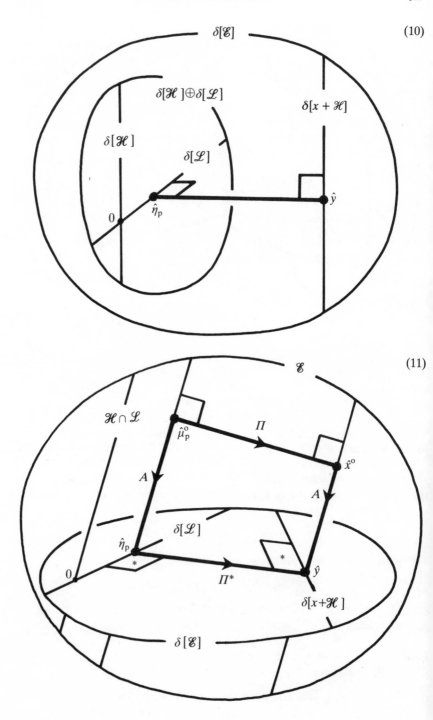

$\mathcal{H} \cap \mathcal{L}$, in which case $(V^\delta)^{-1}$ will be identified with the inner product $(V^\delta)^+$ on $\mathcal{E}$. Let $A$ then denote projection onto $\delta[\mathcal{E}]$ parallel to $\mathcal{H} \cap \mathcal{L}$, and $\Pi$ be $V^{-1}$-orthogonal projection onto $x + \mathcal{H}$. A simple adaptation of Theorem 11 of §11 then gives Fig. 11, in which $\Pi^*$ is $(V^\delta)^+$-orthogonal projection in $\delta[\mathcal{E}]$ onto $\delta[x + \mathcal{H}]$. This shows that $\hat{y}$ is the $(V^\delta)^+$-orthogonal projection of $\hat{\eta}_p$ on $\delta[x + \mathcal{H}]$. Symmetrically, $\hat{\eta}_p$ is the $(V^\delta)^+$-orthogonal projection of $\hat{y}$ on $\delta[\mathcal{L}]$. That $\hat{\eta}_p$, $\hat{y}$ is the only pair of $(V^\delta)^+$-closest vectors follows by the non-singularity of $(V^\delta)^+$ and the disposition of the spaces $\delta[\mathcal{L}]$, $\delta[x + \mathcal{H}]$.                    □

The main result can now be stated:

**Theorem 2**  $\hat{y}$ is the minimum mean square error predictor of $y$, among affine predictors with zero expectation for the error of prediction.

**Proof**  In analogy with (7), a general affine predictor of $y$, based on knowledge of $x + \mathcal{H}$, may be written

$$\bar{y} = a + B\hat{y}.$$

Then

$$E(y - \bar{y}) = -a - (B - 1)\eta,$$

which is identically zero if and only if $a = 0$ and $B$ is an identity on $\delta[\mathcal{L}]$. Then, just as for (8),

$$E(y - \bar{y})^2 = E\{y - \hat{y} + (1 - B)(\hat{y} - \hat{\eta})\}^2 \geqslant E(y - \hat{y})^2, \qquad (12)$$

establishing the optimality of $\hat{y}$.                    □

**Case 3**  Suppose that, in addition to the conditions of the $\mathcal{H} \cap \mathcal{L} = \{0\}$ subcase of Case 2, supplementary information about $\mu$ is available in the form of a completely observed random sample $x_1, \ldots, x_n$ from $P$. A general affine predictor of $x$ may now be written

$$\bar{x}_0 = a + Bx + B_1 x_1 + \ldots + B_n x_n$$

where $B$ must annihilate $\mathcal{H}$. The following 'reduction' lemma allows us to improve on $\bar{x}_0$ immediately.

**Lemma 2**  Let $\hat{\mu}^-$ denote the $V^{-1}$-orthogonal projection of $\bar{x}$ on $\mathcal{L}$, and

$$\bar{x} =_{\text{def}} a + Bx + n\bar{B}\hat{\mu}^- \qquad (13)$$

where $\bar{B} = (B_1 + \ldots + B_n)/n$. Then

(i)  $E(\bar{x}) = E(\bar{x}_0)$,
(ii)  $\text{var}(\bar{x}) \leqslant \text{var}(\bar{x}_0)$,

for all $\mu \in \mathcal{L}$.

**Proof**   (i) is immediate. For (ii), we have

$$\mathrm{var}(B_1 x_1 + \ldots + B_n x_n) = V^{B_1} + \ldots + V^{B_n}$$
$$= V^{B_1 - \bar{B}} + \ldots + V^{B_n - \bar{B}} + nV^{\bar{B}}$$
$$\geqslant nV^{\bar{B}} = \mathrm{var}(n\bar{B}\bar{x}) \geqslant \mathrm{var}(n\bar{B}\hat{\mu}^-)$$

by the Gauss Reduction Theorem of §14.                                    □

So we may restrict our analysis to predictors of type (13), and obtain

**Theorem 3**   The minimum mean square error, affine predictor of $x$ with zero expectation of error is $\hat{x}$, the $V^{*-1}$-orthogonal projection of $\hat{\mu}^-$ on $x + \mathcal{H}$ where

$$V^* = V + \frac{1}{n} V^\Pi \tag{14}$$

with $\Pi = V^{-1}$-orthogonal projection onto $\mathcal{L}$.

**Proof**   For $\bar{x} = a + Bx + C\hat{\mu}^-$, zero expectation of prediction error necessitates

$$a + (B + C)\mu \equiv \mu$$

for $\mu \in \mathcal{L}$. Whence $a = 0$ and $B + C = 1$ on $\mathcal{L}$. Then

$$E(x - \bar{x})^2 = E\{(1 - B)x - (1 - B)\hat{\mu}^-\}^2 = (V^*)^{1-B},$$

where $1 - B$ is an identity on $\mathcal{H}$. The result then follows by the by-now-familiar application of the Corollary of §12.                    □

Figure 15 shows how the two orthogonalities, that of $V^{-1}$ resprsented by □ and that of $V^{*-1}$ represented by ⊞, work together to give $\hat{x}$. The comparison with the case of no additional sample, shown in Fig. 6, is particularly interesting and a little surprising. Note that $\hat{\mu}^-$ is the Gauss estimator based on $x_1, \ldots, x_n$ only. Derivation of the Gauss estimator $\hat{\mu}_{\mathrm{p}}$, based on $x_1, \ldots, x_n$ and what is known of $x$, is left as Exercise 3 of the next section; the associated picture is found to involve a further projection—the $V^{-1}$-orthogonal projection of $\hat{x}$ back onto $\mathcal{L}$!

**Case 4**   When $V$ is unknown in Case 3 and $\mathcal{L} \neq \{0\}$, we may be prepared to use the sample variance

$$S = \frac{1}{n - 1} \sum (x_i - \bar{x})^2,$$

when non-singular, as a plug-in estimator of $V$. The resulting predictor would be given by putting $S$ and $S^* =_{\mathrm{def}} S + \frac{1}{n} S^\Pi$ in place of $V$ and $V^*$,

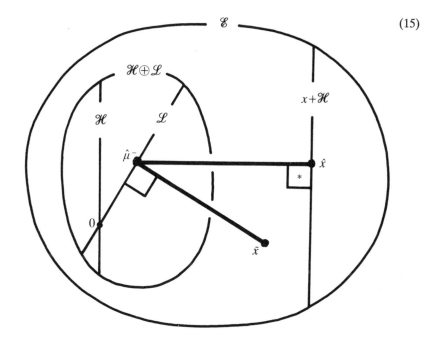

respectively, in the construction of Theorem 3. Then $\hat{x}$ would be the $S^{*-1}$-orthogonal projection on $x + \mathcal{H}$ of the $S^{-1}$-orthogonal projection of $\bar{x}$ on $\mathcal{L}$. However, in view of the randomness of $S$, it is not possible to give $\hat{x}$ theoretical support like that of Theorem 3, and the plug-in predictor must therefore be treated as just one of a range of possibilities.

The problem here appears to be wide open for coordinate-free investigation, but we will here be content to state an optimality of an extension of the plug-in predictor for the special case $\mathcal{L} = \mathcal{E}$. For this case, $S^* = \left(1 + \dfrac{1}{n}\right)S$ and $\hat{\mu}^- = \bar{x}$, so that $\hat{x}$ is the $S^{-1}$-orthogonal projection, $\hat{x}^-$ say, of $\bar{x}$ on $x + \mathcal{H}$. Extend the definition of $\hat{x}^-$ by employing $S^+$-orthogonal projection on $(x + \mathcal{H}) \cap (\bar{x} + \mathcal{R}_S)$ when $S$ is singular. From Theorem 7 of §11 or, more directly, by analogy with eqn (2), we have

$$\hat{x}^- = \bar{x} + \sum (x_i - \bar{x})(\gamma_i - \bar{\gamma})\left\{\sum (\gamma_i - \bar{\gamma})^2\right\}^+ (\gamma - \bar{\gamma}) \qquad (16)$$

where $\gamma = \gamma[x] =_{\text{def}} x + \mathcal{H} \in \mathcal{E}/\mathcal{H}$ and $\gamma_i = \gamma[x_i]$, $i = 1, \ldots, n$. We now exploit the Corollary of §19, in the same way that Theorem 5 of §19 was exploited in the conditional analysis of §22, the conditioning now taking the form $\gamma[x] = \gamma$ and $\gamma[x_i] = \gamma_i$, $i = 1, \ldots, n$. It is also necessary to suppose that the error distribution is linear and homoscedastic in $\mathcal{E}/\mathcal{H}$, and that the predictand $x$ is such that $\gamma$ is in the affine span of $\{\gamma_i\}$. It then follows that $\hat{x}^-$ is the minimum (conditional) variance, affine in $h_1, \ldots, h_n$, conditionally unbiased estimator of $E(x \mid \gamma[x] = \gamma)$. (Here $h_1, \ldots, h_n$ denote the $\mathcal{H}$-components of $x_1, \ldots, x_n$ for any resolution $\mathcal{E} = \mathcal{H} \oplus \mathcal{K}$.) That $\hat{x}^-$ is then also the conditionally optimal predictor of $x$ follows from the conditional independence of $x$ and $\{x_1, \ldots, x_n\}$. Note that the only distributional requirement of the predictand $x$ is that its conditional expectation, given what is known about it, is compatible with the supposed linearity in $\mathcal{E}/\mathcal{H}$.

*Exercises*

1. State and prove a version of Theorem 1 for singular $V$.
2. Accept the invitation in Case 2 to find a proof of the optimality of $\hat{x}$ along the lines suggested following eqn (5).
3. Show that the optimality of $\hat{x}$, established in Case 2 with $\mathcal{H} \cap \mathcal{L} = \{0\}$, may also be proved by replacing the equality in (8) by the identity

$$E(x - \bar{x})^2 = V^{1-B},$$

and then applying Theorem 4 of §12 with $\Pi$ equal to $V^{-1}$-orthogonal projection onto $\mathcal{H} \oplus \mathcal{L}$.
4. Investigate the optimality of $\hat{\mu}_p$ and $\hat{\eta}_p$ in the respective subcases of Case 2.
5. What happens if $x + \mathcal{H}$ intersects $\mathcal{L}$ in Figs 6 or 9?
6. Suppose that, in Case 4, $\mathcal{R}_S + \mathcal{H} = \mathcal{E}$. Show that $\gamma = x + \mathcal{H}$ is then in the affine span of $\{\gamma_i\}$. Use the final part of Exercise 5 of §11 to prove the identity in $v \in \mathcal{V}$

$$[\hat{x}^-, v] = [\bar{x}, v] + [x - \bar{x}, \dot{v}],$$

where $\dot{v}$ is the $S$-orthogonal projection of $v$ on $\mathcal{H}^\square$. (cf. 'best linear predictor', (Dempster 1969, p. 149)).

## §24. Bayesian optimizations

The uncompromising Bayesian approach to the solution of the problems we have been discussing is to have, as it were, 'probabilities on everything', at least on everything that cannot be isolated from the problem in hand. Applied to Gauss's linear model, (1) of §13, this approach might even dispense with the assumption that $V_0$ is known and assign a prior distribution directly to the distribution $P$ of the random $x \in \mathcal{E}$.

When the problem is one of prediction, the Bayesian approach may, at least conceptually, bypass $P$ by directly assigning just enough probability to observations—past and future—to solve the problem. Whatever form it takes, Bayesian method renounces any interest in imposing constraints on estimators or predictors, such as unbiasedness or affine dependence on data. Thus, in more than one respect, Bayesian method goes well beyond the scope of the simple vector space operations we have so far considered.

By 'Bayesian optimizations' for Gauss's linear model, we will mean something quite modest—the elimination of the unbiasedness constraint. This is made possible by the specification of prior knowledge about the mean and variance inner product of $\mu$.

Thus we extend the Gauss model, (1) of §13, to what we will call the *Bayes–Gauss model*:

$$\left.\begin{aligned}
&E(\mu) = 0, \text{ without loss of generality,} \\
&\text{var}(\mu) = U = \bar{\sigma}^2 U_0, \text{ where } U_0 \text{ is known, singular} \\
&\text{or non-singular,} \\
&E(x \mid \mu) = \mu, \\
&\text{var}(x \mid \mu) = V = \sigma^2 V_0, \text{ where } V_0 \text{ is known and} \\
&\text{non-singular,} \\
&\rho = \sigma^2/\bar{\sigma}^2 \text{ is known and } 0 < \rho < \infty.
\end{aligned}\right\} \tag{1}$$

We continue, as before, to be interested only in optimal affine estimation or prediction, but with optimality now defined in terms of unconditional mean square error.

Let us note that $E(\mu) = 0$ and $\text{var}(\mu) = U$ together imply that, with prior probability 1, we have $\mu \in \mathcal{L} = \mathcal{R}_U = \mathcal{R}_{U_0}$, a known linear subspace of $\mathcal{E}$. Thus the previous condition on $\mu$ is subsumed in the present model. We have imposed the condition that $V$ be non-singular to evade the question of the use of the prior knowledge in conjunction with 'singular' information, provided by $x$ about $\mu$, of the sort analysed in §15.

**Theorem 1** The optimal affine estimator of $\mu$ is $\bar{\mu} = U(U + V)^{-1}x$, which has mean square error $U - U(U + V)^{-1}U$.

**Proof** $E(a + Bx - \mu)^2 = E\{a + (B - 1)\mu\}^2 + V^B$
$$= a^2 + U^{B-1} + V^B.$$

Now

$$U^{B-1} + V^B = (B - 1)U(B' - 1) + BVB'$$
$$= \{B - U(U + V)^{-1}\}(U + V)\{B' - (U + V)^{-1}U\}$$
$$+ \{U - U(U + V)^{-1}U\}.$$

Hence the mean square error is minimized if and only if $a = 0$ and $B = U(U + V)^{-1}$, giving the minimized value $U - U(U + V)^{-1}U$. $\qquad\square$

The optimality of $\bar{\mu}$ is inherited by any linear transformation of it. More precisely, if $\gamma = G_1\mu$ where $G_1$ is linear, $\mathscr{E} \to \mathscr{G}$, then $G_1\bar{\mu}$ is the optimal affine estimator of $\gamma$. This can be seen by an extension of the algebraic argument used for Theorem 1. We have, for $\hat{\gamma} = g + Gx$,

$$
\begin{aligned}
E(\hat{\gamma} - \gamma)^2 &= E\{g + (G - G_1)\mu\}^2 + V^G \\
&= g^2 + U^{G-G_1} + V^G \\
&= g^2 + \{G - G_1U(U+V)^{-1}\}(U+V)\{G' - (U+V)^{-1}UG_1'\} \\
&\quad + \{U - U(U+V)^{-1}U\}^{G_1},
\end{aligned}
$$

which is minimized by $g = 0$ and $G = G_1U(U + V)^{-1}$.

We will see in §31 how geometry may be substituted for the pure algebra of these derivations.

There are interesting connections between the Bayesian estimator $\bar{\mu}$ and the Gauss estimator $\hat{\mu}$. These are immediate consequences of some technical properties of the linear transformation

$$
B(\rho) =_{\text{def}} U(U + V)^{-1} = U_0(U_0 + \rho V_0)^{-1}, \qquad \rho > 0.
$$

In the terminology of Theorem 2 of §10, $B(\rho)$ involves an 'augmented dual' of $U_0$.

**Theorem 2**

(i) $\mathcal{N}_{B(\rho)} = \mathcal{R}_U^{V^{-1}}$,

(ii) For $e \in \mathcal{R}_U$, $\lim_{\rho \to 0} B(\rho)e = e$,

(iii) For $e \in \mathcal{R}_U$, $U^+(B(\rho)e, B(\rho)e) \leqslant U^+(e, e)$.

**Proof**  (i) and (ii) are immediate consequences of Theorem 2 (i), (ii), and (iii) of §10 with $I = V_0$, $S = U_0$ and $A(\rho) = (U_0 + \rho V_0)^{-1}$. For (iii), we need to show that

$$
UU^+U \geqslant B(\rho)UU^+UB(\rho)',
$$

which is equivalent to

$$
U \geqslant U(U + V)^{-1}U(U + V)^{-1}U
$$

or

$$
\{U - U(U + V)^{-1}U\} + U\{(U + V)^{-1} - (U + V)^{-1}U(U + V)^{-1}\}U \geqslant 0.
$$

The first term is non-negative by Theorem 1, while the second expression in parentheses is $(U + V)^{-1}V(U + V)^{-1} \geqslant 0$.  □

The following relationships between $\bar{\mu}$ and $\hat{\mu}$, based on Theorem 2, may

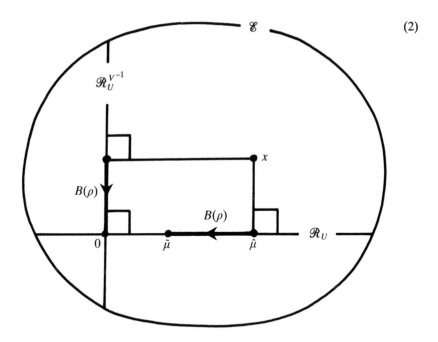

(2)

be readily checked with the aid of Fig. 2.

**Corollary**

  (i)   $\bar{\mu} = U(U + V)^{-1}\hat{\mu}$,

  (ii)  $\lim_{\rho \to 0} \bar{\mu} = \hat{\mu}$,

 (iii)  $U^{+}(\bar{\mu}, \bar{\mu}) \leqslant U^{+}(\hat{\mu}, \hat{\mu})$.

Note that Gauss Reduction is as applicable to the Bayes estimator $\bar{\mu}$ as to any other: Corollary (i) gives the necessary function of $\hat{\mu}$. Corollary (iii), which does not depend on the choice of $U^{+}$, is a coordinate-free statement of a Bayesian shrinkage phenomenon: $\bar{\mu}$ is closer (in terms of $U^{+}$) to the prior mean of $\mu$, than is the optimal Gauss estimator $\hat{\mu}$.

Turning now to Bayesian prediction, we can immediately apply the result on shadow inner product minimization in §12:

**Theorem 3**    The optimal affine predictor, $\bar{x}$, of $x$, given the coset $x + \mathcal{H}$ of $\mathcal{E}/\mathcal{H}$ in which $x$ is known to lie, is the $(U + V)^{-1}$-orthogonal projection of the origin on $x + \mathcal{H}$ and

$$\bar{x} = [1 - H\{H(U + V)^{-1}H\}^{+}H(U + V)^{-1}]\hat{x} \qquad (3)$$

where $H$ is any inner product with $\mathcal{R}_{H} = \mathcal{H}$.

**Proof**   $a + Bx$ is a general affine predictor based on $x + \mathcal{H}$, provided $B$ annihilates $\mathcal{H}$. The mean square error of prediction is

$$E(x - a - Bx)^2 = a^2 + (U + V)^{1-B}. \tag{4}$$

Since $1 - B$ is an identity on $\mathcal{H}$, it follows from the Corollary of §12 with $I = U + V$, $\mathcal{S} = \mathcal{H}$, and $B \to 1 - B$ that $(U + V)^{1-B}$ is minimized when $1 - B$ is $(U + V)^{-1}$-orthogonal projection onto $\mathcal{H}$. With the requirement that $a = 0$ for minimization of (4), this establishes the first part of the theorem. Since $\mathrm{var}(x) = U + V$, the formula (3) for the projection follows from application of Theorem 8 of §11, in conjunction with Fig. 5.     □

$$\tag{5}$$

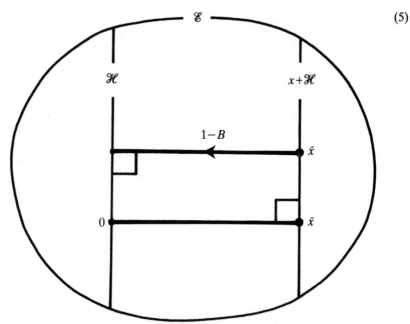

As for estimation so for prediction, the optimality of $\bar{x}$ is inherited by any linear transformations. Thus if $y = G_2 x$ where $G_2$ is linear $\mathcal{E} \to \mathcal{G}$ then $G_2 \bar{x}$ is the optimal predictor of $y$. This may be seen by application of the Transformed Inner Product Theorem of §12. For $\hat{y} = g + Gx$, the mean square error of prediction is $g^2 + (U + V)^{G_2 - G}$. Since $G$ is required only to annihilate $\mathcal{H}$, the sole condition on $G_2 - G$ is that it agree with $G_2$ on $\mathcal{H}$. Theorem 4 of §12 then applies, with $I = U + V$, $\mathcal{S} = \mathcal{H}$, $T = G_2$ restricted to $\mathcal{H}$, and $B = G_2 - G$, to establish the optimality of $G_2 \bar{x}$.

If the direct analogue of Corollary (i) were to apply to this prediction case, the Bayes-optimal affine estimator $\tilde{\mu}_p$, say, would be related to the Gauss estimator $\hat{\mu}_p$ of §23 by $\tilde{\mu}_p = U(U + V)^{-1} \hat{\mu}_p$. The following theorem, uncovered by Wang Jinglong, shows that $\tilde{\mu}_p$ is determined rather by analogy with Corollary (i) in its alternative equivalent form $\bar{\mu} = U(U+V)^{-1}x$.

**Theorem 4** The Bayes-optimal affine estimator of $\mu$ is

$$\bar{\mu}_p = U(U+V)^{-1}\bar{x}. \tag{6}$$

**Proof** Follow the proof of Theorem 1 to show that

$$E(a + Bx - \mu)^2 = a^2 + (U+V)^{B - U(U+V)^{-1}} + \{U - U(U+V)^{-1}U\},$$

whence, for optimality, $a = 0$. The requirement that $B$ annihilate $\mathcal{H}$ is equivalent to $B - U(U+V)^{-1}$ agreeing with $T = -U(U+V)^{-1}$ on $\mathcal{H}$. Theorem 4 of §12 then implies that the optimal $B$ is $U(U+V)^{-1}(1 - \Pi)$, where $\Pi$ is $(U+V)^{-1}$-orthogonal projection onto $\mathcal{H}$, which, with a second use of Fig. 5, delivers $Bx$ as (6). $\square$

It is clear that the validity of Theorems 3 and 4 is unaffected by whether or not $\mathcal{H} \cap \mathcal{L} = \{0\}$. However, the following analogue of Corollary (ii) does require that $\mathcal{H} \cap \mathcal{L} = \{0\}$, without which, as we saw in §23, $\hat{x}$ is not uniquely defined.

**Theorem 5** For the case $\mathcal{H} \cap \mathcal{L} = \{0\}$,

$$\lim_{\rho \to 0} \bar{x} = \hat{x}.$$

**Proof** In the expression (3) for $\bar{x}$, it may be verified that we may take

$$\{H(U+V)^{-1}H\}^+ = H^+(U+V)H^+, \tag{7}$$

provided we choose $\mathcal{R}_{H^+} = (U+V)^{-1}\mathcal{R}_H$ (allowed because $\{(U+V)^{-1}\mathcal{R}_H\} \cap \mathcal{N}_H = \{0\}$). Hence

$$\bar{x} = \hat{x} - HH^+A(\rho)^{-1}H^+HA(\rho)\hat{x},$$

where $A(\rho) = (U_0 + \rho V_0)^{-1}$. Now $\hat{x} - \hat{\mu} \in \mathcal{R}_{U_0}^{V_0^{-1}}$, whence we may apply Theorem 2 (i) of §10 with $S = U_0$, $I = V_0$ to conclude that

$$HA(\rho)(\hat{x} - \hat{\mu}) = \rho^{-1}HV_0^{-1}(\hat{x} - \hat{\mu}) = 0,$$

since $\hat{x} - \hat{\mu}$ is also $V_0^{-1}$-orthogonal to $\mathcal{H}$. So our result will be proved if we can show that

$$\lim_{\rho \to 0}(U_0 + \rho V_0)H^+HA(\rho)\hat{\mu} = 0. \tag{8}$$

By part (iii) of Theorem 2 of §10, $\lim_{\rho \to 0} A(\rho)\hat{\mu}$ exists, while by part (vi)

$$\mathcal{R}_{H^+} = \rho(U_0 + \rho V_0)^{-1}\mathcal{R}_H \to \Pi V_0^{-1}\Pi'\mathcal{R}_H \subset \mathcal{N}_{U_0}.$$

Together these establish (8). $\square$

*Exercises*

1. Prove that, when $\mathcal{H} \cap \mathcal{L} = \{0\}$, $\bar{\mu}_p \rightarrow \hat{\mu}_p$ as $\rho \rightarrow 0$.
2. Verify the assertion about eqn (7).
3. Show that the Gauss estimator $\hat{\mu}_p$ in Case 3 of §23 is

$$\hat{\mu}_p = \hat{\mu}^- + V^{\Pi}(V^{\Pi} + nV)^{-1}(\hat{\mu}^0 - \hat{\mu}^-),\tag{9}$$

where $\hat{\mu}^0$ is the $V^{-1}$-orthogonal projection of $\hat{x}$ on $\mathcal{L}$. (*Hint*: Use the perfecting-of-the-square device of the proof of Theorem 1, the Transformed Inner Product Minimization Theorem of §12, and Theorem 2 of this section.) In what sense does $\hat{\mu}_p$ lie 'between' $\hat{\mu}^-$ and $\hat{\mu}^0$?

# 4
# Generalized inverses

In this chapter, we extend the definition of those 'generalized inverses' of linear transformations that we have already met, in their disguised form as duals of singular inner products. Dispensing with coordinates greatly simplifies the customary treatment of generalized inverses, reducing their multiplicity but preserving their usefulness in the description of theoretical structures and their properties.

## §25. Annihilating and minimal inverses

A dual inner product $I^+$ of a singular inner product $I$ on $\mathcal{V}$ was defined in §11. As a linear transformation from $\mathscr{E}$ to $\mathcal{V}$, $I^+$ acts as an inverse of $I$, when the action of $I$ is restricted to a subspace $\mathcal{U}$ of $\mathcal{V}$ that is complementary to $\mathcal{N}_I$. By its construction, $I^+$ annihilates a subspace of $\mathscr{E}$ complementary to $\mathscr{R}_I$ and, by the symmetry also built into $I^+$, this subspace happens to be $\mathcal{U}^\square$.

The two properties, *restricted inversion* and *annihilation*, characterize what we will call an 'annihilating' inverse of a linear transformation between two general finite-dimensional vector spaces $\mathscr{X}$, $\mathscr{Y}$.

**Definition 1** An *annihilating inverse, $A^+$,* of a linear transformation $A$, $\mathscr{X} \to \mathscr{Y}$, is a linear transformation, $\mathscr{Y} \to \mathscr{X}$, that

(i) inverts $A$, $\mathscr{A} \to \mathscr{R}_A$, where $\mathscr{A}$ is some subspace of $\mathscr{X}$ complementary to $\mathcal{N}_A$,
(ii) annihilates all vectors in $\mathscr{B}$, some subspace of $\mathscr{Y}$ complementary to $\mathscr{R}_A$.

The two components of this definition are shown in Fig. 1. In some applications of the concept of a generalized inverse, the annihilation property may not be required—for example, if the inverse is required to act only on $\mathscr{R}_A$.

**Definition 2** A *minimal inverse, $A^-$,* of $A$, $\mathscr{X} \to \mathscr{Y}$, is a linear transformation, $\mathscr{Y} \to \mathscr{X}$, that inverts $A$, $\mathscr{A} \to \mathscr{R}_A$, where $\mathscr{A}$ is some subspace of $\mathscr{X}$ complementary to $\mathcal{N}_A$.

The picture, Fig. 2, for a minimal inverse is naturally simpler than Fig. 1. The following two theorems state algebraic characterizations of $A^-$ and $A^+$. As we will see, the proofs are surprisingly tedious. This difficulty

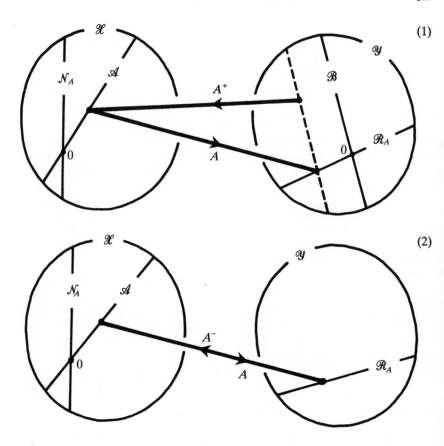

in translating from one mathematical structure to an equivalent one is, of course, symmetrical in character. It does not, in itself, suggest that either structure is to be preferred. It is noteworthy, however, that the geometrical structure has separately identifiable features which are, in a sense, compounded in the one or two algebraic identities that are their equivalent. Perhaps the problems of translation have something to do with this compounding.

**Theorem 1**   $A^-$ is a minimal inverse if and only if

$$AA^-A = A.$$  (3)

**Proof**   If $A^-$ is a minimal inverse then, for $x \in \mathcal{N}_A$, $AA^-Ax = 0 = Ax$, while, for $x \in \mathcal{A}$, $AA^-Ax = Ax$. Hence (3) obtains.

Conversely, if (3) holds then $A^-$ cannot annihilate any non-zero vector in $\mathcal{R}_A$, and, moreover, $(A^-\mathcal{R}_A) \cap \mathcal{N}_A = \{0\}$. Then $\mathcal{A} = A^-\mathcal{R}_A$ has the necessary properties for $A^-$ to be minimal.    □

**Theorem 2**   $A^+$ is an annihilating inverse if and only if $AA^+A = A$ and

$$A^+AA^+ = A^+. \tag{4}$$

**Proof**   If $A^+$ is annihilating, we have $AA^+A = A$ from Theorem 1. Moreover, for $y \in \mathcal{R}_A$, $A^+AA^+y = A^+y$ while, for $y \in \mathcal{B}$, $A^+AA^+y = 0 = A^+y$. Hence (4) also holds.

Conversely, if $AA^+A = A$ and (4) obtain then we have the minimal property as in Theorem 1. Moreover, (4) implies $\dim \mathcal{R}_{A^+} \leqslant \dim \mathcal{R}_A$ or $\dim \mathcal{N}_{A^+} \geqslant \dim \mathcal{Y} - \dim \mathcal{R}_A$. As for $A^-$, $A^+$ cannot annihilate any non-zero vector in $\mathcal{R}_A$, whence $A^+$ must annihilate a subspace complementary to $\mathcal{R}_A$, which serves as $\mathcal{B}$.   □

For our coordinate-free purposes, $A^-$ and $A^+$ are the only types of generalized inverse that need to be considered. The greater profusion of types to be found in the standard coordinatized treatments arises from investing one or both of the spaces $\mathcal{X}$ and $\mathcal{Y}$ with an inner product. Thus the renowned Moore–Penrose generalized inverse is, from our point of view, the particular choice of $A^+$ in which $\mathcal{A}$ is $I$-orthogonal to $\mathcal{N}_A$ and $\mathcal{B}$ is $J$-orthogonal to $\mathcal{R}_A$, where $I$ and $J$ are the inner products corresponding to orthogonality of the coordinate bases of $\mathcal{X}$ and $\mathcal{Y}$ respectively. The two conditions, that need to be adjoined to the identities $AA^+A = A$, $A^+AA^+ = A^+$ to give a purely algebraic characterization of the Moore–Penrose inverse, are those that make the projections, $A^+A$ and $AA^+$, $I$-symmetric and $J$-symmetric respectively. In duality terms, these conditions are

$$IA^+A = (A^+A)'I, \tag{5}$$

$$JAA^+ = (AA^+)'J, \tag{6}$$

corresponding to the geometrical conditions of orthogonality for the subspaces $\mathcal{A}$ and $\mathcal{B}$, respectively. The picture for $A^+$ is shown in Fig. 7.

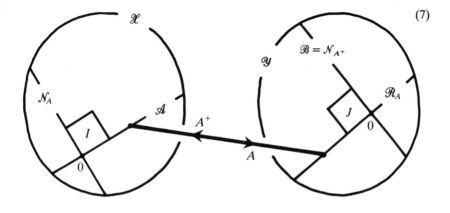

(7)

Distinctions involving inner products are clearly of importance when the theory is given practical realization in any application. However, such distinctions may be unhelpful if they are introduced at too early a stage in the theoretical development.

We end this section by using minimal generalized inverses to meet the promise of §23 (Case 2, $\mathscr{H} \cap \mathscr{L} = \{0\}$), by deriving explicit formulae for the estimator $\hat{\mu}_p$ and predictor $\hat{x}$.

**Theorem 3**  For the case $\mathscr{H} \cap \mathscr{L} = \{0\}$ with inner product dualtors $H$, $U$ such that $\mathscr{R}_H = \mathscr{H}$, $\mathscr{R}_U = \mathscr{L}$, we have

$$\left. \begin{aligned} \hat{x} &= \{1 - HH^-V(H^-H + U^-U)V^{-1}\}x, \\ \hat{\mu}_p &= UU^-V(H^-H + U^-U)V^{-1}x, \end{aligned} \right\} \tag{8}$$

where $H^-$, $U^-$ are minimal generalized inverses of $H$ and $U$, chosen to annihilate $\mathscr{L}$, $\mathscr{H}$, respectively, and such that

$$V^{-1}(\mathscr{H} \oplus \mathscr{L}) = \mathscr{R}_{H^-} \oplus \mathscr{R}_{U^-}.$$

**Proof**  By Theorem 3 of §6, the subspace $V^{-1}(\mathscr{H} \oplus \mathscr{L})$ or $V^{-1}\mathscr{H} \oplus V^{-1}\mathscr{L}$, which has dimension complementary to that of $(\mathscr{H} \oplus \mathscr{L})^{\square}$ or $\mathscr{H}^{\square} \cap \mathscr{L}^{\square}$, will be actually complementary to the latter. Hence $\mathscr{L}^{\square} \cap V^{-1}(\mathscr{H} \oplus \mathscr{L})$ is complementary in $\mathscr{L}^{\square}$ to $\mathscr{H}^{\square} \cap \mathscr{L}^{\square}$, and also complemen-

(9)

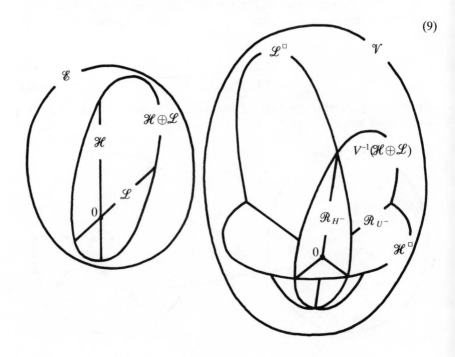

tary in $\mathcal{V}$ to $\mathcal{H}^{\square}$. So we may choose $H^-$ with

$$\mathcal{R}_{H^-} = \mathcal{L}^{\square} \cap V^{-1}(\mathcal{H} \oplus \mathcal{L})$$

and, likewise, $U^-$ with

$$\mathcal{R}_{U^-} = \mathcal{H}^{\square} \cap V^{-1}(\mathcal{H} \oplus \mathcal{L}).$$

These choices uniquely satisfy the required conditions. Figure 9 provides useful reassurance. Now $\hat{x}$ and $\hat{\mu}_p$ are characterized by the equations

$$\left.\begin{aligned} HV^{-1}(\hat{x} - \hat{\mu}_p) &= 0, \\ UV^{-1}(\hat{x} - \hat{\mu}_p) &= 0. \end{aligned}\right\} \tag{10}$$

Writing $\hat{x} = \hat{h} + k$, we may allocate $u$, $v$ in $\mathcal{V}$ so that $\hat{h} = Hu$, $\hat{\mu}_p = Uv$. Then (10) may be written

$$\binom{H}{U} V^{-1}(H\ U)\binom{u}{-v} = -\binom{H}{U} V^{-1}k. \tag{11}$$

With the stated conditions on $H^-$ and $U^-$, it may be verified that minimal inverses of $\binom{H}{U}$, $(H\ U)$ are $(H^-\ U^-)$, $\binom{H^-}{U^-}$, respectively, and that (11) has a solution

$$\binom{u}{-v} = -\binom{H^-}{U^-} V (H^- U^-)\binom{H}{U} V^{-1}k,$$

whence

$$\hat{h} = -HH^- V(H^- H + U^- U)V^{-1}k,$$
$$\hat{\mu}_p = UU^- V(H^- H + U^- U)V^{-1}k.$$

These give (8) with $k$ instead of $x$ on the right-hand side. The replacement by $x$ is permitted because the operators annihilate $\mathcal{H}$.  □


*Exercises*

1. Extending the applicability of Theorem 1 of §24 to the case when $U + V$ is singular, show that $U^{B-1} + V^B$ is minimized when $B = U(U + V)^-$ where the superscript $-$ denotes any minimal inverse.
2. (Continuation) Show that the minimizing $B$ annihilates $\mathcal{C}$, the $V^+$-orthogonal complement of $\mathcal{R}_U \cap \mathcal{R}_V$ in $\mathcal{R}_V$, and that $(U + V)^- \mathcal{R}_U$ is complementary to $\mathcal{N}_U$. (*Hint:* For $c \in \mathcal{C}$ and $v = (U + V)^- c$, show that $Uv \in \mathcal{R}_U \cap \mathcal{R}_V$, and then that $V^+(c, Uv) = U(v, v) + V^+(Uv, Uv)$.)

## §26. Duals of generalized inverses

For $A^+$ as shown in Fig. 1, the dual transformation $(A^+)'$, $\mathcal{X}' \to \mathcal{Y}'$, has range space $(\mathcal{N}_{A^+})^\square = \mathcal{B}^\square$ and null space $(\mathcal{R}_{A^+})^\square = \mathcal{A}^\square$. These identities justify the lower half of Fig. 1. In this picture, the possibility has been raised that $(A^+)'$ does not reverse $A'$ between $\mathcal{B}^\square$ and $\mathcal{R}_{A'}$. This possibility is represented by the inequality $y'_1 \neq y'_2$. However, from the relationship of $(A^+)'$ and $A^+$,

$$[x', x] = [y'_1, y],$$

while, from that of $A'$ and $A$,

$$[x', x] = [y'_2, y]$$

So $[y'_1 - y'_2, y] \equiv 0$ identically in $y \in \mathcal{R}_A$. Additionally, $[y'_1 - y'_2, b] \equiv 0$ identically in $b \in \mathcal{B}$. Together, these identities imply that $y_1 = y_2$ i.e. that $(A^+)'$ does in fact satisfy both of the conditions for it to be an annihilating generalized inverse of $A'$. Hence the commutation $(A^+)' = (A')^+$ is justified.

This geometric proof is perhaps only a little longer than the algebraic

(1)

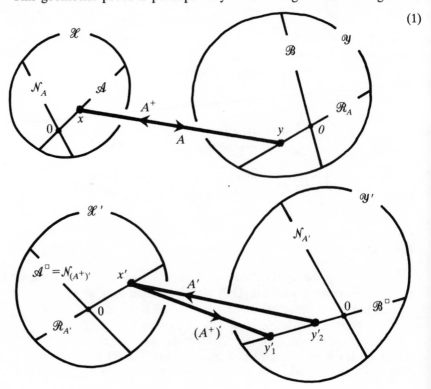

procedure of introducing inner products $I$ and $J$, as if for the Moore–Penrose inverse, and then verifying that the four algebraic conditions for commutability are satisfied.

However, for the proof that $(A^-)'$ is a minimal generalized inverse of $A'$, there seems little doubt that the advantage lies with the algebraic proof

$$AA^-A = A \Rightarrow A'A^{-\prime}A' = A',$$

which appears to leave geometry at the starting gate!

*Exercise*

1. If $\mathcal{Y} = \mathcal{X}'$ and $A$ is symmetric, show that $A^+$ is also symmetric if and only if $\mathcal{A}$ and $\mathcal{B}$ are chosen to be biorthogonal complements.

## §27. Review

Before introducing some representations of generalized inverses that will be needed in the final chapter, let us review some unacknowledged uses in earlier sections of the concept of an annihilating inverse.

The first of these was with the clairvoyantly notated $V^+$ in eqn (2) of §11. As a linear transformation, $V^+$ is an annihilating inverse of $V$, with range $\mathcal{W}$ and null space $\mathcal{W}^\square$. The necessary connection here between range and null space (which are independently assignable for a general annihilating inverse) is a consequence of the symmetry imposed on $V^+$ by the mode of construction.

The same observations apply to $I^+$ as to $V^+$. However, in the dualtor $(\int t^2 \, d\alpha)^+$ of eqn (9) in §11, we find an application of annihilating inverses that is strictly unnecessary—in the mathematical sense. For $t_1, t_2 \in \mathcal{T} = \mathcal{R}_{(\int t^2 d\alpha)}$, we have

$$\left( \int t^2 \, d\alpha \right)^+ (t_1, t_2) = \left[ t_1, \left( \int t^2 \, d\alpha \right)^+ t_2 \right] = \left[ t_1, \left( \int t^2 \, d\alpha \right)^- t_2 \right], \quad (1)$$

where $(\int t^2 \, d\alpha)^+$, acting as linear transformation on $t_2$, has been replaced by any minimal inverse $(\int t^2 \, d\alpha)^-$. The arbitrariness in the latter does not affect the evaluation with $t_1$. This reduction to minimality enters eqn (9) of §11 thus:

$$\left( \int st \, d\alpha \right) \left( \int t^2 \, d\alpha \right)^+ t = \int s(e) \left[ t(e), \left( \int t^2 \, d\alpha \right)^+ t \right] d\alpha(e)$$

$$= \int s(e) \left[ t(e), \left( \int t^2 \, d\alpha \right)^- t \right] d\alpha(e)$$

$$= \left( \int st \, d\alpha \right) \left( \int t^2 \, d\alpha \right)^- t.$$

A similar argument may be used in the Π of eqn (12) in §11, allowing the replacement of the annihilating $+$ by the minimal $-$, as well as for the associated variance formulae such as var($\hat{\mu}$) in §14.

*Exercise*

1. Extend the definition of $[I, J]$ in §9 so that $I^+$ may be replaced by $I^-$ in Theorem 9 of §11.

## §28. Representations

Even a brief inspection of the standard texts on generalized inverses reveals an impressive range of relationships between superficially different algebraic formulae. The following theorems provide coordinate-free statements and proofs for a small selection of the more useful of these relationships. But firstly let us state a useful technical result, a special case of which appeared in the proof of Theorem 8 of §11. Throughout this section, $A$ is linear $\mathcal{X} \to \mathcal{Y}$ and $I$, $J$ are non-singular inner products on $\mathcal{X}$, $\mathcal{Y}$, respectively.

(1)

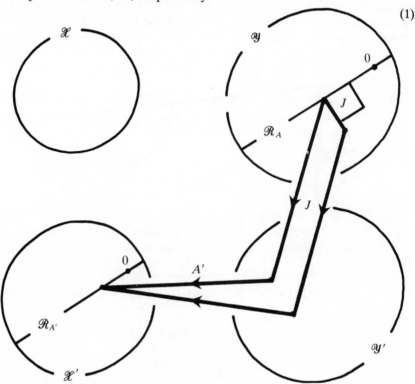

**Theorem 1**

$$\mathcal{N}_{A'J} = \mathcal{R}_A^J.$$

**Proof**  The result follows from composition of $J\mathcal{R}_A^J = \mathcal{R}_A^\square$ (Theorem 2 of §6) and $\mathcal{R}_A^\square = \mathcal{N}_{A'}$ (extending Theorem 1 of §5 to a general linear transformation), together with dimensionality considerations.    □

Figure 1 for Theorem 1 reveals a pronounced 'robot arm' feature that will be put to good use in several of the ensuing theorems. The reader may recall the previous appearance of the feature, unadvertized, in Fig. 13 of §11.

**Theorem 2**

$$A^+ =_{\text{def}} (A'JA)^- A'J$$

is an annihilating inverse of $A$ with $\mathcal{N}_{A^+} = \mathcal{R}_A^J$, whatever the choice of minimal inverse.

**Proof**  Adding a few items to (1) gives us Fig. 2. Clearly $QA'J$ has the properties required for $A^+$. The recognition that $Q$, extended arbitrarily

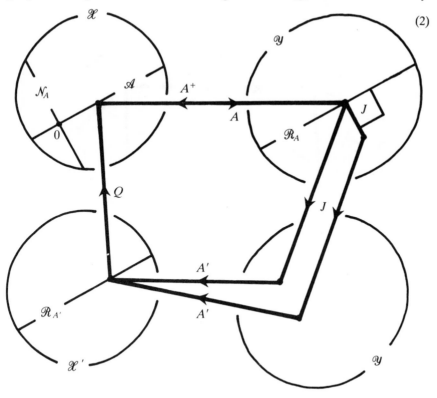

(2)

to be linear $\mathscr{X}' \to \mathscr{X}$, is a minimal inverse of $A'JA$ then gives the result.                                                                                    □

**Corollary**   $AA^+ = A(A'JA)^-A'J$ is $J$-orthogonal projection onto $\mathscr{R}_A$.

**Theorem 3**

$$A^+ =_{\text{def}} I^{-1}A'(AI^{-1}A')^+$$

is an annihilating inverse with

$$\mathscr{N}_{A^+} = \mathscr{N}_{(AI^{-1}A')^+}$$

and

$$\mathscr{R}_{A^+} = \mathscr{N}_A^I.$$

**Proof**   The picture is now as depicted in Fig. 3, which is built up by initial arbitrary choices of $\mathscr{N}_Q$, complementary to $\mathscr{R}_A$, and of $\mathscr{R}_Q$, complementary to $\mathscr{N}_{A'}$. The action of $Q$ on $\mathscr{R}_A$, however, is then fixed by the requirement that it is the inverse of $AI^{-1}A'$ between $\mathscr{R}_Q$ and $\mathscr{R}_A$.

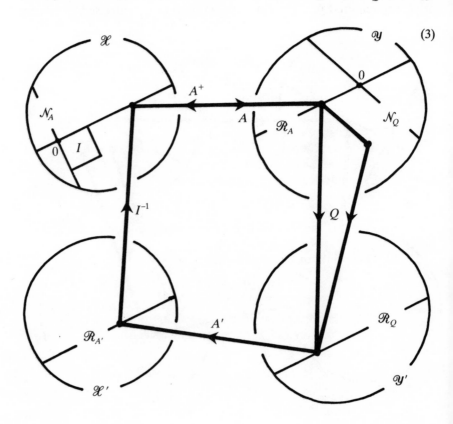

(3)

Thus $Q = (AI^{-1}A')^+$ if we choose

$$\mathcal{N}_{(AI^{-1}A')^+} = \mathcal{N}_Q.$$

Finally, observe (literally) that $I^{-1}A'Q$ then satisfies the requirements for $A^+$.                                                                                  □

**Corollary**   $A^+A = I^{-1}A'(AI^{-1}A')^+A$ is $I$-orthogonal projection onto $\mathcal{N}_A^I$.

**Theorem 4**   If $\mathscr{Z}$ is finite-dimensional, $\mathscr{X} = \mathcal{N}_A \oplus \mathcal{N}_B$ where $B$ is linear, $\mathscr{X} \to \mathscr{Z}$, and $K$ is a non-singular inner product on $\mathscr{Z}$, then

$$A^+ =_{\text{def}} (A'JA + B'KB)^{-1}A'J$$

is an annihilating inverse with $\mathcal{N}_{A^+} = \mathcal{R}_A^J$ and $\mathcal{R}_{A^+} = \mathcal{N}_B$.

**Proof**   (See Fig. 4, which is a six-wheeler with robotics!) We start with $x \in \mathscr{X}$, resolve it into its $\mathcal{N}_A$ and $\mathcal{N}_B$ components, and send these by the transformations $B$ and $A$, respectively, on roundabout journeys to $\mathcal{R}_{B'}$ and $\mathcal{R}_{A'}$, respectively, in $\mathscr{X}'$ where they are reconstituted to make $x'$.

Algebraically,

$$x' = (A'JA + B'KB)x,$$

or $x = Qx'$ with

$$Q = (A'JA + B'KB)^{-1}.$$

The action of $QA'J$ on $y \in \mathscr{Y}$ satisfies the requirements for $A^+$.         □

*Exercises*

1. You are given that $A$, linear $\mathscr{X} \to \mathscr{Y}$, equals $BC$, where $C$ is linear, $\mathscr{X} \to \mathscr{Z}$, with $\mathcal{R}_C = \mathscr{Z}$, and that $B$ is linear, $\mathscr{Z} \to \mathscr{Y}$, with $\mathcal{N}_B = \{0\}$. Make a geometrical analysis of the transformation

$$I^{-1}C'(CI^{-1}C')^{-1}(B'JB)^{-1}B'J,$$

and show that it is the annihilating inverse of $A$ with null space $\mathcal{R}_A^J$ and range $\mathcal{N}_A^I$.

2. Use Theorem 2 of §10 to show that

$$\lim_{\rho \to 0} (A'JA + \rho I)^{-1}A'J$$

is also the annihilating inverse of $A$ with null space $\mathcal{R}_A^J$ and range $\mathcal{N}_A^I$.

3. The following exercises explore versions of Theorem 2 that evade the use of dual spaces.

(i)   Show, without dualities, that the identity in $x \in \mathscr{X}$ and $y \in \mathscr{Y}$,

$$I(x, A^T y) \equiv J(Ax, y),$$

(4)

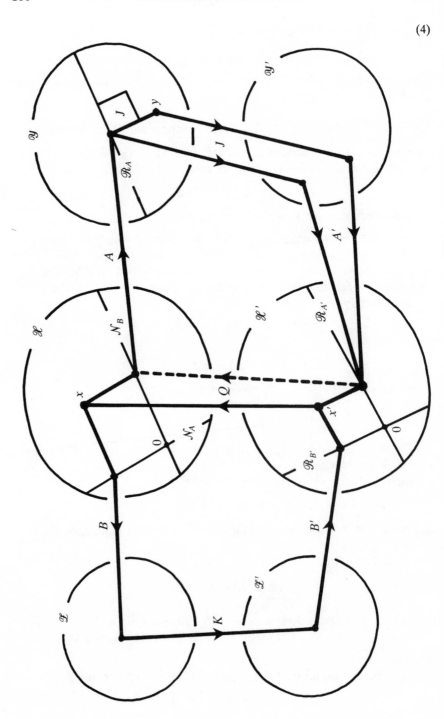

defines a linear transformation $A^T$, $\mathcal{Y} \to \mathcal{X}$, with $\mathcal{N}_{A^T} = \mathcal{R}_A^J$ and $\mathcal{R}_{A^T} = \mathcal{N}_A^I$.

(ii) Using only the spaces $\mathcal{X}$ and $\mathcal{Y}$, draw the figures representing the identity

$$A^+ = (A^T A)^- A^T$$

for the cases (a) $\mathcal{A}(=\mathcal{R}_{A^+})$ different from $\mathcal{N}_A^I$ and (b) $\mathcal{A} = \mathcal{N}_A^I$. What is special about $A^+$ for case (a) and for case (b)?

# 5
# Parametrizations

In this chapter, we will strengthen the connections between the coordinate-free approach and the more traditional accounts of the topics under study. This will be done by extending the uses already made of a third vector space, and exploring the possible forms of linkage between $\mathscr{E}$ (with its dual $\mathscr{V}$) and a supplementary parameter vector space, together with its dual.

## §29. Underparameters and overparameters

In the versions, so far considered, of the Gauss model

$$\left. \begin{array}{l} E(x) = \mu \in \mathscr{L}, \text{ a subspace of } \mathscr{E}, \\ \mathrm{var}(x) = V, \end{array} \right\} \tag{1}$$

the estimation of $\mu$, or of a marginal parameter $\tau[\mu]$ (§14), does not take us outside $\mathscr{E}$ and $\mathscr{V}$. As already indicated in Chap. III, it is sometimes the case that such estimation is merely a stage towards the estimation of a related *parameter of interest*. Our present task is to develop some more coordinate-free machinery for this. We will suppose that $V$ is known up to proportionality.

Two types of relationship will be considered, separately. The first type is one in which the parameter of interest, $\theta$, is a vector in a separate vector space $\Theta$, called *parameter space*, with $\theta$ related to $\mu$ by a known linear transformation from $\mathscr{L}$ onto $\Theta$. The second differs from this only in that the direction of the known linear transformation is from $\Theta$ onto $\mathscr{L}$.

The cases of interest are when these transformations, $\mathscr{L} \to \Theta$ or $\Theta \to \mathscr{L}$, are not invertible. If they were invertible, we would simply be involved in a relabelling of the vectors in $\mathscr{L}$. From the coordinate-free viewpoint, such transformations to an isomorphic vector space are of slight interest, even though they can be of appreciable value in the coordinatization needed for numerical calculations.

**Definition 1** If $\theta = T\mu$, where $T$ is a known linear transformation, $\mathscr{L} \to \Theta$, with $\mathscr{R}_T = \Theta$, then $\theta$ will be called an *underparameter* of the model (1).

**Definition 2** If $\mu = X\theta$, where $X$ is a known linear transformation,

102

(2)

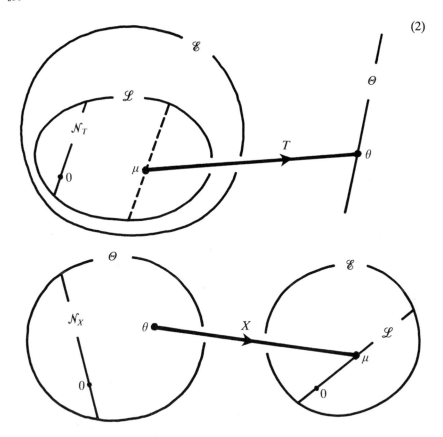

$\Theta \to \mathscr{L}$, with $\mathscr{R}_X = \mathscr{L}$, then $\theta$ will be called an *overparameter* of the model (1).

The reason for the terminology is obvious. In Definition 1, the parameter of interest, $\theta$, underparametrises $\mathscr{L}$ in the sense that, given non-invertibility of $T$, there are distinct values of $\mu \in \mathscr{L}$ with the same value of $\theta$. In Definition 2, however, the reverse is the case, and it is $\mathscr{L}$ that is overparametrised by $\theta$. Figure 2 shows both types of parametrization: for underparametrization, the vector space $\Theta$ has been flattened in order to emphasize the 'under-' aspect. An underparametrization may clearly be viewed as a relabelling of a marginal parameter in $\mathscr{L}$.

## §30. Gauss estimation

For an analysis of underparameter estimation, the work has already been done in §§14–15. For the case of $V$ non-singular, we simply put $\mathscr{G} = \Theta$, $\gamma = \theta$ and $G_1 = T$ in Theorem 3 of §14, and conclude that $\hat{\theta} = T\hat{\mu}$ is the

(1)

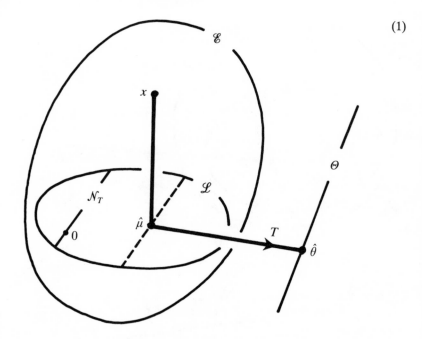

minimum variance, affine, unbiased estimator of $\theta$. For $V$ singular, the analogue of Theorem 3 could have been stated in §15, leading to the same conclusion for $\hat{\theta} = T\hat{\mu}$, with 'sub-affine' in place of 'affine'.

Figure 1 covers both the non-singular and singular cases. It illustrates the pivotal role of the Gauss estimator $\hat{\mu}$, the $V^{-1}$- or $V^+$-orthogonal projection of $x$ onto $\mathscr{L}$ or $\phi[x] \cap \mathscr{L}$, respectively.

There is a fundamental triviality at the core of our second problem— Gauss estimation for the overparametrized case. It is one that is not easily concealed, even in the purely algebraic approach, and we will make little effort to do so.

Consider firstly the case of non-singular $V$. If attention is, from the start, restricted to affine estimators of $\theta$ of the form $a + Bx$, where $a \in \Theta$ and $B$ is linear, $\mathscr{E} \rightarrow \Theta$, then the Gauss Reduction of §14 justifies further restriction to the class

$$\{a + B\hat{\mu} : a \in \Theta, \ B \text{ linear } \mathscr{E} \rightarrow \Theta\}, \tag{2}$$

where $\hat{\mu}$ is the Gauss estimator of $\mu$.

**Definition** The *null bias set*

$$\Theta_{a,B} =_{\text{def}} \{\theta : E(a + B\hat{\mu} \mid \theta) = \theta\}$$

is the set of values of $\theta \in \Theta$ for which $a + B\hat{\mu}$ is an unbiased estimator.

**Lemma 1**   $\Theta_{a,B}$ is an affine manifold, the translate of some subspace of $\Theta$.

**Proof**   For scalar $\lambda$, if $\theta'$ and $\theta''$ are in $\Theta_{a,B}$ then

$$E(a + B\hat{\mu} \mid \theta = \lambda\theta' + (1-\lambda)\theta'') = a + BX(\lambda\theta' + (1-\lambda)\theta'')$$
$$= \lambda(a + BX\theta') + (1-\lambda)(a + BX\theta'') = \lambda\theta' + (1-\lambda)\theta''.$$

Hence $\lambda\theta' + (1-\lambda)\theta'' \in \Theta_{a,B}$.      $\square$

**Lemma 2**   Translates of $\mathcal{N}_X$ intersect $\Theta_{a,B}$ in at most one point.

**Proof**   If not, we would have $\theta'$, $\theta''$ $(\theta' \neq \theta'')$ in $\Theta_{a,B} \cap (\mathcal{N}_X - \phi)$ for some $\phi$ in $\Theta$. Hence

$$a + BX\theta' = \theta' \tag{3}$$

$$a + BX\theta'' = \theta'', \tag{4}$$

whence $0 = BX[(\phi + \theta') - (\phi + \theta'')] = \theta' - \theta''$, a contradiction.      $\square$

Lemmas 1 and 2 imply that the choices of $a$, $B$ with $a \in \mathcal{N}_X$ and $B$ some minimal generalized inverse, $X^-$, give the maximal null bias sets, i.e. any $\Theta_{a,B}$ is included in at least one such set.

If we eschew any reduction of variance that may be obtained by choosing $B$ not equal to any $X^-$ and require maximality of the null bias set, we have then arrived at a further reduction of the choice of estimator to the class

$$\{n + X^-\hat{\mu} : n \in \mathcal{N}_X, \quad X^- \text{ a minimal inverse}\}. \tag{5}$$

Observe that $n + X^-\hat{\mu}$ may also be expressed in the form $n + X^+x$ where $X^+$ annihilates $\mathcal{L}^{V^{-1}}$ and $\mathcal{R}_{X^+} = \mathcal{R}_{X^-}$.

The route to the class (5) may be criticized as an attempt to provide optimal estimators for too many values of $\theta$ at the same time. To avoid any illusion of grandeur, it may be preferable to start afresh by defining the marginal parameter, $\nu$, by

$$\nu = \nu[\theta] = \theta + \mathcal{N}_X \in \Theta/\mathcal{N}_X.$$

The vector space $\Theta/\mathcal{N}_X$ is in 1–1 correspondence with $\mathcal{L}$. Writing $\nu = T\mu$, there is a reassuring consensus about the estimation of $\nu$ among the estimators in (5):

$$\hat{\nu} = \nu[n + X^-\hat{\mu}] = X^-\hat{\mu} + \mathcal{N}_X = T\hat{\mu}. \tag{6}$$

More cogent in favour of $T\hat{\mu}$ is the fact that, by §14, it is the minimum variance, affine, unbiased estimator of the marginal parameter $\nu$.

The picture, Fig. 7, for optimal overparameter estimation in the case of non-singular variance, incorporates all the points and relationships we

(7)

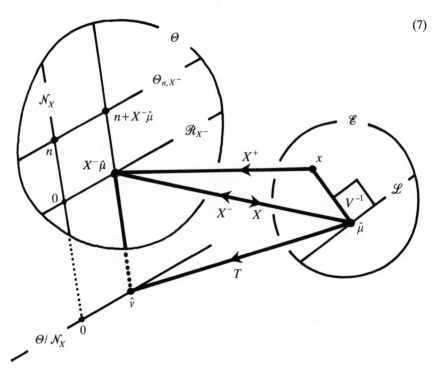

have established. Finally, we may use Theorem 2 of §28 to write

$$X^-\hat{\mu} = (X'V^{-1}X)^-X'V^{-1}X \qquad (8)$$

where $(X'V^{-1}X)^-$ is any minimal inverse.

Now for the somewhat less straightforward singular case. By reviewing the arguments of §15 and applying them to the overparametrization problem, it is fairly clear, without going into detail, that the class of estimators (5), or the more restrained estimator $\hat{v}$ in (6), continues to provide optimal Gauss estimation. Now, of course, $\hat{\mu}$ is the $V^+$- orthogonal projection of $x$ onto $\phi[x] \cap \mathcal{L}$, the set of values of $\mu$ consistent with $x$. Picture (7), however, still serves for the singular case.

The only item that has to be modified to take account of the singularity of $V$ is the alternative formula for $X^-\hat{\mu}$ given by eqn (8). Some delicacy is needed to obtain a valid reformulation. The problem is most easily resolved if we explicitly involve the dual vector spaces $\mathcal{V}$ and $\Theta'$, and, for this, it helps if we start with a picture that builds on (7). To reduce visual clutter, we will not reproduce all elements of (7)—only those needed to support the dual space development. However, to accommodate the features of the singular variance case, it is necessary to expand the representation of $\mathcal{L} = \mathcal{R}_X$.

With a little trial and error, we arrive at Fig. 9. At first glance, Fig. 9 may well appear horrendously complex. However, apart from the introduction of a subspace $\mathcal{S}^+$ of $\mathcal{F}$ and an as yet unspecified linear transformation $Q$, $\mathcal{E} \to \mathcal{V}$, all of the elements of (9) should prove familiar on further inspection. The subspace $\mathcal{S}^+ \subset \mathcal{F}$ is the $V^+$-orthogonal complement in $\mathcal{F}$ of $\mathcal{S} = \mathcal{F} \cap \mathcal{L}$. Since $\hat{\mu}$ is the $V^+$-orthogonal projection of $x$ on $\phi[x] \cap \mathcal{L}$, it follows that $x - \hat{\mu} \in \mathcal{S}^+$.

The transformation $Q$ is required to supply the upper-arm of the crucial robot arm pictured. As it stands, Fig. 9 is a statement of possibility rather than feasibility. In effect, it asks whether we can find a useful transformation $Q$ with the properties portrayed:

(i)  $Q\mathcal{S}^+ \subset \mathcal{N}_{X'}$ so that $X'Q$ annihilates $\mathcal{S}^+$,
(ii) $Q\mathcal{L}$ is complementary to $\mathcal{N}_{X'}$, so that $X'Q$ has range $\mathcal{R}_{X'}$.

(9)

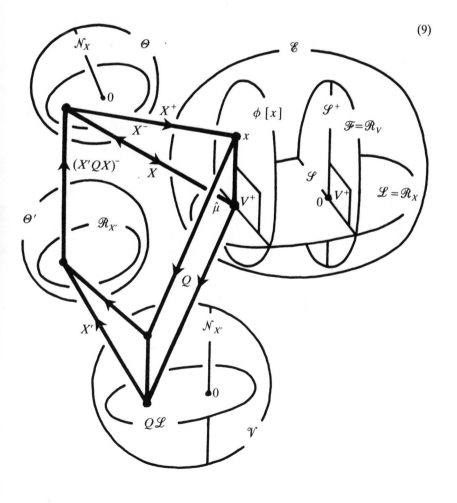

Noting that $\mathscr{S}^+ \cap \mathscr{L} = \{0\}$, and that $\dim \mathscr{L} = \dim \mathscr{R}_X$ is also the dimension of any subspace of $\mathscr{V}$ complementary to $\mathscr{N}_{X'}$, the *existence* of such $Q$ is not in doubt. The emphasis must be on finding a *useful* one. The search is, happily, easily satisfied.

The requirements (i) and (ii) on $Q$ are reminiscent of Exercise 2 of §25. Setting $U = XWX'$, where $W$ is any non-singular inner product on $\Theta'$, we have $\mathscr{R}_U = \mathscr{R}_X = \mathscr{L}$ and $\mathscr{N}_U = \mathscr{N}_{X'}$. It then follows from Exercise 2 that the conditions (i) and (ii) are satisfied by $Q = (XWX' + V)^-$, any minimal inverse.

Then, from Fig. 9, we see that

$$X^-\hat{\mu} = (X'QX)^-X'Qx$$
$$= \{X'(XWX' + V)^-X\}^-X'(XWX' + V)^-x. \tag{10}$$

*Exercises*

1. By eqn (8), $V^{-1}$-orthogonal projection onto $\mathscr{L}$ may be expressed as the transformation

$$X(X'V^{-1}X)^-X'V^{-1}. \tag{11}$$

However, by Theorem 8 of §11, the same projection may also be expressed as the transformation

$$XWX'(XWX'V^{-1}XWX')^-XWX'V^{-1}$$

where $W$ is any non-singular inner product on $\Theta'$. (We have here taken up the option, noted at the end of §27, of using any $-$ instead of $+$ for the generalized inverse involved.)

Construct a geometric reconciliation of these different formulae based on transformations between $\Theta$, $\mathscr{E}$ and their duals. (*Hint*: It is necessary only to show that $WX'(XWX'V^{-1}XWX')^-XW$ is a minimal inverse of $X'V^{-1}X$, for which proof two copies of $\mathscr{E}$ and of $V$ are helpful. A simpler linkage with §11 is provided by changing Theorem 8 of §11 so that $U$ is replaced by $X$, with (13) of §11 then based on $\mathscr{E}$ and $\Theta$ and their duals. The change delivers expression (11) by the same mode of proof as for Theorem 8 of §11, which did not depend on $U$ being an inner product.)

2. For the singular case displayed in Fig. 9, show that $V\mathscr{N}_{X'} = \mathscr{S}^+$ and $V\mathscr{N}_{X'}^V = \mathscr{S}$. (*Hint*: Use Theorem 2 of §6 with $\mathscr{U} = \mathscr{N}_{X'}$ and $I = V$, and Theorem 6 of §11 with $I = V$.)

3. (Continuation) Show that there exist minimal inverses $V^-$ with $V^-\mathscr{S}^+ \subset \mathscr{N}_{X'}$ and $V^-\mathscr{L}$ complementary to $\mathscr{N}_{X'}$, so that such inverses may be used as the required transformation $Q$.

4. (Continuation) Show that $V^-$, satisfying the required conditions,

may be chosen to be symmetric and that then

$$\text{var}(\hat{\mu}) = X(X'V^-X)^-X'.$$

5. Show that $Q = (XWX' + V)^-$ cannot be a minimal inverse of $V$, unless $\mathcal{R}_X \cap \mathcal{R}_V = \{0\}$.

## §31. Bayesian estimation

In §24, we described a Bayesian analysis for the Gauss model that did not make use of any explicit parametrization of the subspace of means $\mathcal{L}$. The core of the analysis was located entirely within $\mathcal{E}$.

For a Bayesian treatment of the parametrized case, whether it be over or under-parametrized, it is also convenient to work within a single vector space. For then we can exploit the results of Chap. III on optimal projection and avoid any recourse to purely algebraic method. The product

$$\mathcal{B} =_{\text{def}} \Theta \times \mathcal{E}$$

provides the necessary space. For the dual of $\mathcal{B}$, we use $\mathcal{B}' = \Theta' \times \mathcal{V}$, in which the dual $\Theta'$ of $\Theta$ plays the role of biorthogonal complement of $\mathcal{E}$ while $\mathcal{V}$ acts as biorthogonal complement of $\Theta$. When there is no risk of confusion, we may identify the vectors $(\theta, 0)$ and $(0, x)$ in $\mathcal{B}$ with the vectors $\theta$ and $x$ respectively, and identify operators such as $(\theta, 0)^2$ and $(\theta, 0)(0, x)$ with $\theta^2$ and $\theta x$, respectively. Furthermore, it is useful, when there is no risk of confusion, to be able to treat $\mathcal{B}$ as $\Theta \oplus \mathcal{E}$ and to write $b = \theta + x$.

The Bayesianized Gauss model, (1) of §24, applies directly to the underparametrized case (Definition 1 of §29). Writing $b = (\theta, x)$, we have $E(b) = 0$ and $\text{var}(b) = \tilde{V}$, say, where

$$\tilde{V} = \begin{pmatrix} TUT' & TU \\ UT' & U+V \end{pmatrix}. \tag{1}$$

The partitioning of $\tilde{V}$ here corresponds to the component spaces $(\Theta, \mathcal{E})$ of $\mathcal{B}$. As linear transformation, $\mathcal{B}' \to \mathcal{B}$, its action on vectors $b' = (\theta', v)$ in $\mathcal{B}'$ satisfies

$$\tilde{V}b' = (TUT'\theta' + TUv, \ UT'\theta' + (U+V)v), \tag{2}$$

in accordance with the usual matrix conventions. However, all suggestion of matrix method could be avoided by exploiting the above-mentioned identifications and adopting the representation

$$\tilde{V} = E(T\mu + x)^2,$$

at least for anything involving $\tilde{V}$.

Theorem 1 of §23 is now applicable with $\mathscr{E}$, $\mathscr{H}$, $\mathscr{K}$, $V$ replaced by the current $\mathscr{B}$, $\Theta$, $\mathscr{E}$, $\tilde{V}$, respectively. The conclusion is that the optimal affine estimator of $\theta$ is, with justifiable avoidance of generalized inverse notation,

$$TU(U+V)^{-1}x, \qquad (3)$$

which equals $TU(U+V)^{-1}\hat{\mu}$ by the Corollary of §24. Expression (3) also equals $T\bar{\mu}$ where $\bar{\mu}$ is the Bayes estimator of $\mu$ found in §24 by purely algebraic argument. The geometric derivation here seems to fit in better with our overall approach.

For the overparametrized case, Definition 2 of §29, with $\mu = X\theta$, it is necessary to provide a prior mean and variance of $\theta$. Suppose, without loss of generality, that $E(\theta) = 0$ and $\text{var}(\theta) = W$. In this case,

$$\tilde{V} = \begin{pmatrix} W & WX' \\ XW & XWX' + V \end{pmatrix}. \qquad (4)$$

Theorem 1 of §23 now applies to give the optimal affine estimator

$$\begin{aligned} \tilde{\theta} &= WX'(XWX' + V)^{-1}x \\ &= WX'(XWX' + V)^{-1}\hat{\mu}, \end{aligned} \qquad (5)$$

where the final step in (5) follows from the non-singular analogue of the results established for $Q$ at the end of §30 or, by more general argument, from the Gauss Reduction Theorem of §14.

There is an alternative form of $\tilde{\theta}$ that is rather a curiosity when derived by purely algebraic argument, but that has an interesting, geometrically constructive derivation.

**Theorem**

$$\begin{aligned} \tilde{\theta} &= (WX'V^{-1}X + 1)^{-1}WX'V^{-1}x \\ &= (WX'V^{-1}X + 1)^{-1}WX'V^{-1}\hat{\mu}. \end{aligned} \qquad (6)$$

**Proof** The transformation $X'V^{-1}$, $\mathscr{L} \to \mathscr{R}_{X'} \subset \Theta'$, is 1–1. Hence $\tilde{\theta}$ is equivalently the optimal affine transformation of $X'V^{-1}\hat{\mu}$ which, by Gauss Reduction, must also be the optimal transformation of $X'V^{-1}x$. Forming $\mathscr{C} = \Theta \times \Theta'$ and writing

$$c = (\theta, X'V^{-1}x)$$

we have $E(c) = 0$ and $\text{var}(c) = \tilde{V}_c$ where

$$\tilde{V}_c = \begin{pmatrix} W & WX'V^{-1}X \\ X'V^{-1}XW & X'V^{-1}(XWX' + V)V^{-1}X \end{pmatrix}.$$

Then, following Theorem 1 of §23 with the relaxation to minimal inverse sanctioned as in §27 and to singularity of $\bar{V}_c$ (§23, Exercise 1),

$$\begin{aligned}
\bar{\theta} &= WX'V^{-1}X\{X'V^{-1}(XWX'+V)V^{-1}X\}^-X'V^{-1}x \\
&= WX'V^{-1}X(WX'V^{-1}X+1)^{-1}(X'V^{-1}X)^-X'V^{-1}x \\
&= WX'V^{-1}X\{1-(WX'V^{-1}X+1)^{-1}WX'V^{-1}X\}(X'V^{-1}X)^-X'V^{-1}x \\
&= WX'V^{-1}x - WX'V^{-1}X(WX'V^{-1}X+1)^{-1}WX'V^{-1}x \\
&= (WX'V^{-1}X+1)^{-1}WX'V^{-1}x = (WX'V^{-1}X+1)^{-1}WX'V^{-1}\hat{\mu}. \quad (7)
\end{aligned}$$

(The non-singularity invoked for $WX'V^{-1}X+1$, equivalently for $X'V^{-1}XW+1$, holds because

$$(X'V^{-1}XW+1)\theta' = 0 \Rightarrow [X'V^{-1}XW\theta', W\theta'] + [\theta', W\theta'] = 0$$
$$\Rightarrow [\theta', W\theta'] = 0 \Rightarrow W\theta' = 0 \Rightarrow \theta' = 0, \text{ by the supposition.}$$

The final step in (7) holds because $X'V^{-1}$ annihilates $x - \hat{\mu}$.)          □

*Exercises*

1. Show that, given $E(\theta) = 0$ and $\text{var}(\theta) = W$, the choice $n = 0$ in (5) of §30 is optimal in terms of unconditional mean square error.
2. Prove that, when $W$ is non-singular (in the overparametrized case),

   (i)  the estimator $X^-x$ minimizes the expected squared bias if $\mathscr{R}_{X^-}$ is $W^{-1}$-orthogonal to $\mathscr{N}_X$ (cf. Chipman 1964),
   (ii) among such estimators, $\hat{\theta} = X^-\hat{\mu}$ minimizes variance.

3. Show that, when $W$ is non-singular, $\bar{\theta}$ in eqn (5) is $W^{-1}$-orthogonal to $\mathscr{N}_X$, and that $\bar{\theta} = X^-\bar{\mu}$ if $\mathscr{R}_{X^-}$ is $W^{-1}$-orthogonal to $\mathscr{N}_X$, where $\bar{\mu}$ is defined as in §24, Theorem 1, with $U = XWX'$.
4. Develop Exercise 2 for the case when $W$ is singular.
5. Show that

$$\bar{\theta} = WX'V^{-1}X(WX'V^{-1}X+1)^{-1}\hat{\theta}$$

where $\hat{\theta} = X^-\hat{\mu}$ without condition on $\mathscr{R}_{X^-}$. (*Hint*: Examine line 2 of (7).)
6. (Continuation) Prove the following inequalities that compare the magnitudes of the deviations of $\bar{\theta}$ and $\hat{\theta}$ from the average of the prior mean 0 and the least-squares estimate $\hat{\theta}$:

   (i)  $X'V^{-1}X(\bar{\theta} - \tfrac{1}{2}\hat{\theta}, \bar{\theta} - \tfrac{1}{2}\hat{\theta}) \leqslant X'V^{-1}X(\tfrac{1}{2}\hat{\theta}, \tfrac{1}{2}\hat{\theta})$
   (ii) If $\mathscr{R}_{X'} \subset \mathscr{R}_{W^+}$ and $\mathscr{R}_X \subset \mathscr{R}_W$ then

$$W^+(\bar{\theta} - \tfrac{1}{2}\hat{\theta}, \bar{\theta} - \tfrac{1}{2}\hat{\theta}) \leqslant W^+(\tfrac{1}{2}\hat{\theta}, \tfrac{1}{2}\hat{\theta})$$

(cf. Chamberlain and Leamer 1976).

### §32. Updating for the Kalman filter

Consider now a general problem of optimal affine Bayes estimation, in which data $x \in \mathscr{E}$ for estimation of $\theta \in \Theta$, a finite-dimensional vector space, is acquired in two stages. At first, the only available data corresponds, as in §23, to the knowledge of $x + \mathscr{H}$, where $\mathscr{H}$ is a subspace of $\mathscr{E}$, or, equivalently, to the knowledge of $k$, the component of $x$ in some subspace $\mathscr{K}$ complementary to $\mathscr{H}$. The optimal affine estimator based on this knowledge is $\tilde{\theta}_0$, say. We will exhibit the 'updating' relationship between $\tilde{\theta}_0$ and $\tilde{\theta}$, where $\tilde{\theta}$ is the optimal affine estimator based on the full data $x$ that becomes available in the second stage.

Suppose that the only additional information is the prior mean and variance (known up to proportionality) of $b = (\theta, x)$. Suppose, without loss of generality, that $E(b) = 0$ and write $\bar{V} = \text{var}(b)$, supposed non-singular. The updating relationship is given by:

**Theorem 1**

$$\tilde{\theta} = \tilde{\theta}_0 + \text{cov}(\theta, \tilde{x})\text{var}(\tilde{x})^{-}\tilde{x}, \tag{1}$$

where

$$\tilde{x} = x - \hat{x} \tag{2}$$

with $\hat{x} = \text{cov}(x, k)\text{var}(k)^{-}k$.

**Proof** Extending the method of the previous section, we draw the necessary picture, Fig. 3, and invoke Theorem 11 of §11. The right-angles $\square$, $\boxtimes$ in Fig. 3 denote $\bar{V}^{-1}$ and $\text{var}(x)^{+}$-orthogonal projections, respectively. Theorem 11 of §11 applies with $\mathscr{E}$, $\mathscr{R}$, $\mathscr{S}$, $\mathscr{A}$, $I$ changed to $\mathscr{B}$, $\mathscr{E}$, $x + \mathscr{H} \oplus \Theta$, $x + \Theta$, $\bar{V}$, respectively, and with its origin regarded as at $x$ in $\mathscr{E}$ (without loss of generality). The trapezium in Fig. 15 of §11, on which the identity (14) of §11 finds expression, is here reduced to the triangle in Fig. 3 standing on the vector $\overline{0\hat{x}}$. The identity justifies the labelling $\tilde{\theta}_0 \in \Theta$ for the vector joining the two projections shown in the triangle. The form for $\hat{x}$ derives from eqn (2) of §23. It can then be seen that the 'updating' adjustment $\tilde{\theta} - \tilde{\theta}_0$ is minus the $\bar{V}^{-1}$-orthogonal projection of $\tilde{x}$ on $\Theta$. But the $\bar{V}^{-1}$-orthogonal projection of $b$ on $\mathscr{H} \oplus \Theta$ is $\tilde{x} + \theta - \tilde{\theta}_0$. Hence the $\bar{V}^{-1}$-orthogonal projection of $\tilde{x}$ on $\Theta$ is, equivalently (see Exercise 9 of §11), its $\text{var}(\tilde{x} + \theta - \tilde{\theta}_0)^{+}$-orthogonal projection. From Theorem 7 of §11 with $\mathscr{S} = \Theta$ and $\mathscr{T} = \mathscr{H}$, this is $\text{cov}(\tilde{\theta}_0 - \theta, \tilde{x})\text{var}(\tilde{x})^{-}\tilde{x}$. The fact that $\text{cov}(\tilde{\theta}_0, \tilde{x}) = 0$ (Exercise 1 of this section) then gives eqn 1. □

There is a specialization of eqn (1) that will have its most important application in a geometrical treatment of the Kalman filter (Luenberger

(3)

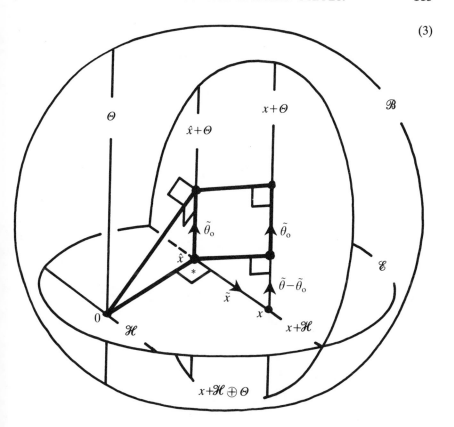

1969). From the process of specialization, it is possible to isolate a preliminary theorem that is of some independent, general interest.

**Theorem 2** (*Reduction*)  Writing $x = h + k$, where $h \in \mathcal{H}$, $k \in \mathcal{K}$, suppose that $h = X\theta + g$ where $X$ is fixed linear, $\Theta \to \mathcal{H}$, and $g \in \mathcal{H}$ is uncorrelated with $k$ and $\theta$. Then the optimal affine estimator of $\theta$, may be found in the class of estimators affine in $(\bar{\theta}_0, h)$.

(We describe this as a 'reduction' theorem in recognition of the substantive case for which the affine transformation $k \to \bar{\theta}_0$ is not invertible.)

**Proof**  A little effort is needed to construct an economical picture of what is going on here, but, once this is done, the result will be seen almost literally to fall out. (The lines of an alternative, mainly algebraic proof are given in an exercise. This application of the two approaches provides an excellent comparison of their respective merits.) In the first step of picture-building, we create a separate vector space $\mathcal{G}$ to

(4)

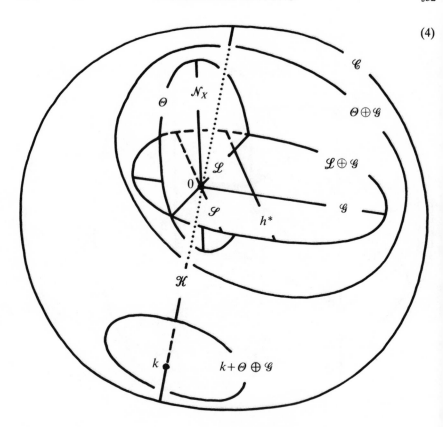

accommodate the random 'error' vector $g$. The space

$$\mathscr{C} =_{\text{def}} \mathscr{K} \oplus \Theta \oplus \mathscr{G}$$

will then hold everything we need. The second-stage data vector $h$ is a function of $\theta$ and $g$, and is built into Fig. 4 by the use of natural correspondence and isomorphism. The first-stage data vector $k$ has to be put into 'higher dimensions': the crucial action will take place in $k + \Theta \oplus G$ and will be analysed in a separate figure. The first natural isomorphism is between $\{\mu = X\theta : \theta \in \Theta\}$ and any subspace, $\mathscr{L}$, of $\Theta$ complementary to $\mathscr{N}_X$. The addition operation in the equation $h = X\theta + g$ refers to that in a common vector space, the $\mathscr{H}$ of the set-up. Using this form of addition, temporarily, we define $\mathscr{S}$ to be the subspace of $\mathscr{L} \oplus \mathscr{G}$ such that the components $\theta$ and $g$ satisfy

$$X\theta + g = 0,$$

and then make $\mathscr{H}$ isomorphic to $(\mathscr{L} \oplus \mathscr{G})/\mathscr{S}$.

(5)

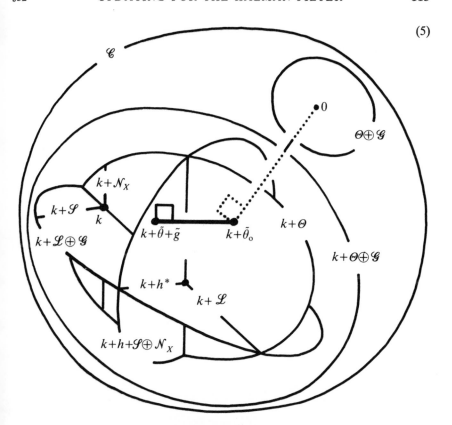

Figure 4 shows the second-stage data vector $h$ as $h^* = h + \mathscr{S}$, one of the cosets in this vector space.

Now transfer attention to Fig. 5, which relegates $\Theta \oplus G$ to the background, and shows the two $\tilde{V}^{-1}$-orthogonal projections involved in our proof, now focussed on $k + \Theta \oplus \mathscr{G}$. The crucial vectors in Fig. 5 are $k + \tilde{\theta}_0$, the first-stage $\tilde{V}^{-1}$-orthogonal projection of 0 on $k + \Theta \oplus \mathscr{G}$, and $k + \tilde{\theta} + \tilde{g}$, the second-stage projection of $k + \tilde{\theta}_0$ on $k + h + \mathscr{S} \oplus \mathscr{N}_X$. The fact that the $\tilde{V}^{-1}$-orthogonal projection of 0 on $k + \Theta \oplus \mathscr{G}$ is $k + \tilde{\theta}_0$ with $\mathscr{G}$-component zero is a consequence of the $\tilde{V}^{-1}$-orthogonality of $\mathscr{G}$ to $\Theta \oplus \mathscr{H}$. In the second stage, $h = X\tilde{\theta} + \tilde{g}$.

The proof is completed by the trivial observation that the $\Theta$-component, $\tilde{\theta}$, of the second-stage projection is an affine transformation of $\tilde{\theta}_0$ and $h$.                    □

The value of Theorem 2 lies in its statement of sufficient conditions under which a possibly high-dimensional data vector $k$ may be reduced to $\tilde{\theta}_0$, in preparation for updating $\tilde{\theta}_0$ by a second data vector $h$. The form of the updating is given by the following theorem.

**Theorem 3**   (*Kalman filter updating*) Under the conditions of this section and of Theorem 2, the updated optimal affine estimator is

$$\tilde{\theta} = \tilde{\theta}_0 + \Sigma_0 X'(X\Sigma_0 X' + R)^-(h - X\tilde{\theta}_0) \qquad (6)$$

where $\Sigma_0 = E(\tilde{\theta}_0 - \theta)^2$ and $R = E(g^2)$. The inner product $\Sigma_0$ is updated by

$$\Sigma = E(\tilde{\theta} - \theta)^2 = \Sigma_0 - \Sigma_0 X'(X\Sigma_0 X' + R)^- X\Sigma_0. \qquad (7)$$

**Proof**   We have $\tilde{\theta}_0 = \theta_0(k)$ say. Let $\tilde{\Theta} = \{\theta_0(k) : k \in \mathcal{K}\}$. By Theorem 2, the optimal estimator is affine in $(\tilde{\theta}_0, h)$. We may therefore apply Theorem 1, with $\mathcal{K}$ replaced by $\tilde{\Theta}_0$ and $\mathcal{E}$ by $\mathcal{Y} = \mathcal{H} \oplus \tilde{\Theta}_0$, to give

$$\tilde{\theta} = \tilde{\theta}_0 + \text{cov}(\theta, \bar{y})\text{var}(\bar{y})^- \bar{y} \qquad (8)$$

where

$$\bar{y} = h - \text{cov}(h, \tilde{\theta}_0)\text{var}(\tilde{\theta}_0)^- \tilde{\theta}_0$$
$$= h - X \text{cov}(\theta, \tilde{\theta}_0)\text{var}(\tilde{\theta}_0)^- \tilde{\theta}_0.$$

The minimization of $E(\tilde{\theta}_0 - \theta)^2$ by the affine-in-$k$ $\tilde{\theta}_0$ implies that $\text{cov}(\theta, \tilde{\theta}_0) = \text{var}(\tilde{\theta}_0)$, whence $\bar{y} = h - X\tilde{\theta}_0$. It follows that, in eqn (8),

$$\text{cov}(\theta, \bar{y}) = \text{cov}\{\theta, X(\theta - \tilde{\theta}_0)\} = \text{cov}\{\theta - \tilde{\theta}_0, X(\theta - \tilde{\theta}_0)\} = \Sigma_0 X'$$

and

$$\text{var}(\bar{y}) = \text{var}\{X(\theta - \tilde{\theta}_0) + g\} = X\Sigma_0 X' + R,$$

establishing eqn (6). Then from (8), using $\text{cov}(\tilde{\theta}_0, \bar{y}) = 0$,

$$E(\tilde{\theta} - \theta)^2 = E(\tilde{\theta}_0 - \theta)^2 - \text{cov}(\theta, \bar{y})\text{var}(\bar{y})^-\text{cov}(\bar{y}, \theta)$$

which equals (7).                                                              □

*Exercises*

1. Show that $\text{cov}(\tilde{\theta}_0, \bar{x}) = 0$ (for the proof of Theorem 1). (*Hint*: $\hat{x}$ and $\bar{x}$ are uncorrelated, and $\tilde{\theta}_0$ is a linear transformation of $\hat{x}$.)

2. Devise an algebraic proof of Theorem 2. (One such proof starts by supposing $\tilde{\theta}_0 = a_0 + B_0 k$, writing a general affine-in-$(h, k)$ estimator as $\tilde{\theta} = a + B_1 h + B_2 k$, and then showing that the 'reduced' estimator

$$\tilde{\theta}_r =_{\text{def}} a + B_1 h + B_2 B_0^- B_0 k,$$

which is affine in $(\tilde{\theta}_0, h)$, has the same expectation as and no larger variance than $\tilde{\theta}$.)

# References

Chamberlain, G. and Leamer, E. E. (1976). Matrix weighted averages and posterior bounds. *J. Roy. Statist. Soc. B* **38**, 73–84.

Chipman, J. S. (1964). On least squares with insufficient observations. *J. Amer. Statist. Assoc.* **59**, 1078–1111.

Dempster, A. P. (1969). *Elements of Continuous Multivariate Analysis*. Reading, Mass.: Addison-Wesley Publishing Co..

Drygas, H. (1970). *The Coordinate-free Approach to Gauss–Markov Estimation.* (*Lecture Notes in Operations Research and Mathematical Systems 40.*) Berlin: Springer–Verlag.

Eaton, M. L. (1983). *Multivariate Statistics: A Vector Space Approach.* New York: Wiley.

Eaton, M. L. (1978). A note on the Gauss–Markov theorem, *Ann. Inst. Statist. Math.* A **30**, 181–184.

Gauss, C. F. (1823). Theoria Combinationis Observationum Erroribus Minimis Obnoxiae. *Werke.* **4**, 1–93.

Halmos, P. R. (1958). *Finite-dimensional Vector Spaces* (2nd edn). Princeton: D. Van Nostrand Co., Inc.

Kruskal, W. H. (1968). When are Gauss–Markov and least squares estimators the same? A coordinate-free approach. *Ann. Math. Statist.* **39**, 70–75.

Kruskal, W. H. (1975). The geometry of generalised inverses. *J. Roy. Statist. Soc. B* **37**, 272–283.

Luenberger, D. G. (1969). *Optimization by Vector Space Methods.* New York: Wiley.

Neuwirth, E. (1982). Parametric deviations in linear models. *Probability and Statistical Inference.* Ed. W. Grossman *et al.* pp. 257–264. Boston: D. Reidel Publishing Co.

Rao, C. R. (1945). Information and accuracy attainable in the estimation of statistical parameters. *Bull. Calcutta Math. Soc.* **37**, 81–91.

Rao, C. R. and Yanai, H. (1985). Generalized inverse of linear transformations: a geometric approach. *Lin. Alg. and Its Applns.* **66**, 87–98.

Stone, M. (1977). A unified approach to coordinate-free multivariate analysis. *Ann. Inst. Statist. Math.* A **29**, 43–57.

Stone, M. (1979). A pictorial treatment of generalized inverses for statistics based on dual vector spaces. *Math. Operationsforsch. Statist. ser. Statistics.* **10**, 3–17.

# Glossary

| | |
|---|---|
| $\alpha, \beta, \rho, \sigma$ | real scalars |
| $e, v, \ldots$ | vectors |
| $\mu$ | vector expectation |
| $A, B, \Pi, \ldots$ | linear transformations |
| $1$ | identity transformation |
| $I, J, \ldots$ | inner products and associated linear transformations |
| $\leqslant$ | partial ordering of inner products |
| $U, V, \ldots$ | variance inner products and associated linear transformations |
| $\mathcal{V}$ | variable space |
| $\mathcal{E}$ | evaluator space |
| $[\,,\,]$ | value function |
| $\mathcal{H}, \mathcal{K}, \mathcal{L}$ | subspaces of $\mathcal{E}$ |
| $\mathcal{U}, \mathcal{W}$ | subspaces of $\mathcal{V}$ |
| $\mathcal{F}$ | error subspace |
| $e + \mathcal{H}$ | $\{e + h : h \in \mathcal{H}\}$, a coset of $\mathcal{H}$ |
| $\mathcal{E}/\mathcal{H}$ | vector space of cosets of $\mathcal{H}$ in $\mathcal{E}$ |
| $(.)'$ | dual of a vector space or linear transformation |
| $(.)^{\square}$ | bi-orthogonal complement of a subspace |
| $(.)^I$ | $I$-orthogonal complement of a subspace |
| $\mathcal{N}_{(.)}$ | null space of linear transformation |
| $\mathcal{R}_{(.)}$ | range space of a linear transformation |
| $\oplus$ | direct sum of two vector spaces |
| $(\hat{\cdot}), (\check{\cdot}), (\bar{\cdot})$ | estimators or predictors of vectors |
| $(.)^A$ | shadow or transformed inner product |
| $(.)^{\mathcal{F}}$ | $\mathcal{F}$-marginal inner product |
| $\tau[e]$ | coset of form $e + \mathcal{F}$ |
| $(.)^+$ | a dual inner product for a singular inner product and an annihilating inverse |
| $(.)^-$ | minimal inverse |
| $-\,-\,-\,-\,-\,-$ | vector space hidden from viewer by intervening items |
| $\ldots\ldots$ | vector space in higher dimensions than those pictured |
| $\square$ | orthogonality |

# Index

affine estimation 5
affine prediction 5
agreement of two transformations on a
    subspace 41
annihilating generalized inverse 89
associated linear transformation 13
augmented dual inner product 27

Bayes–Gauss model 83
Bayesian optimization 83
Bayesian shrinkage 85
bilinearity 12
bi-orthogonal complement 9

conditional estimation 65
conditional unbiasedness 66
coset of a subspace 5
covariance of two ordered random
    evaluators 21

dual error distribution 22, 30
dual inner product 15, 31
dual linear transformation 10, 26
dual vector space 3
dualtor 19

equivalent variances 50
error distribution linear in 65
error distribution homoscedastic in 67
error estimation 48
error subspace 30
error vector 5
evaluator 2
evaluator space 3
expectation of a random evaluator 5

Gauss estimator 44
Gauss Optimality 46
Gauss Reduction 44
Gauss's linear model 43
generalized inverses 89

homoscedasticity 67

inner product ($I$) 12
$I$-isometry 16
$I$-orthogonal complement 13
$I$-orthogonal projection 15
$I$-orthogonal subspaces 14
$I$-symmetry 16

Kalman filter updating 116
Kronecker-type matrix 56

least-squares manifold 61
least-squares (multivariate) 54
linear model 43
location model 59
$\mathscr{L}$-equivalent variances 50

marginal inner product 26
marginal parameter 45, 105
mean square error 44
minimal generalized inverse 89
minimum inner product 40
Moore–Penrose inverse 91
multivariate least-squares 52

non-negative linear transformation 11
non-singular inner product 12
null-bias set 104

orthogonal complement 13
orthogonal subspaces 14
orthogonal projection 15
Orthogonal Shadow Inequality 41
overparameter 103

partial observation 7, 72
partial ordering of inner products 40
plug-in estimator 69, 80
positive linear transformation 11